普通高等教育机械类专业教材

互换性与测量技术基础
（双语）

万一品　贾　洁　主　编
宋绪丁　主　审

人民交通出版社股份有限公司
北　京

Abstract
内容提要

This book is an English-Chinese Bilingual teaching material for mechanical specialty in general higher education. The book expounds the interchangeability of mechanical parts and the basic knowledge of measurement technology, and introduces the latest standards of tolerance and fit in China. The book consists of six chapters: introduction, size tolerances and fits, geometric tolerances, surface roughness, fundation of geometrical quantity measurement and tolerances and fits for typical parts. Each chapter of the book is attached with relevant exercises to meet the needs of teaching.

This book is suitable for the teaching of mechanical courses in universities, for English-Chinese bilingual teaching materials, and can also be used as a reference book for mechanical engineers.

本书是普通高等教育机械类专业英汉双语教材。全书阐述了机械零部件的互换性和测量技术的基本知识，介绍了我国公差与配合方面的最新标准。全书包括绪论、尺寸公差与配合、几何公差、表面粗糙度、几何量测量基础和典型零部件的公差与配合，共6章。书中各章附有相关习题，以配合教学的需要。

本书适用于高等院校机械类的专业课程教学，适合作为英汉双语教材，也可以作为机械工程人员的参考用书。

图书在版编目(CIP)数据

互换性与测量技术基础:汉文、英文/万一品，贾洁主编. —北京:人民交通出版社股份有限公司，2021.12
　ISBN 978-7-114-17227-4

Ⅰ.①互… Ⅱ.①万… ②贾… Ⅲ.①零部件—互换性—高等学校—教材—汉、英②零部件—测量技术—高等学校—教材—汉、英 Ⅳ.①TG801

中国版本图书馆 CIP 数据核字(2021)第 133852 号

Huhuanxing yu Celiang Jishu Jichu(Shuangyu)
书　　名：互换性与测量技术基础(双语)
著 作 者：万一品　贾　洁
责任编辑：钟　伟
责任校对：孙国靖　魏佳宁
责任印制：张　凯
出版发行：人民交通出版社股份有限公司
地　　址：(100011)北京市朝阳区安定门外外馆斜街 3 号
网　　址：http://www.ccpcl.com.cn
销售电话：(010)59757973
总 经 销：人民交通出版社股份有限公司发行部
经　　销：各地新华书店
印　　刷：北京交通印务有限公司
开　　本：787×1092　1/16
印　　张：9.5
字　　数：216 千
版　　次：2021 年 12 月　第 1 版
印　　次：2021 年 12 月　第 1 次印刷
书　　号：ISBN 978-7-114-17227-4
定　　价：30.00 元

(有印刷、装订质量问题的图书由本公司负责调换)

PREFACE

"Interchangeability and Foundation of Measurement Technology" is one of the main technical basic courses required for mechanical instrumentation and electromechanical majors in colleges and universities. It is also a basic subject closely related to the development of the machinery industry. As Chinese education gradually goes to the world, it is urgent to compile textbooks with Chinese characteristics and in line with international standards. Therefore, with the support of Chang'an University and China Communications Press Co., Ltd., the editors began to try to compile basic bilingual textbooks on interchangeability and measurement technology.

According to the characteristics of fewer teaching hours, this textbook strengthens the basic concepts, basic theories, basic methods and basic skills, and compiles the contents according to the latest national and international standards.

1. The compilation of the Chinese content of the textbook

(1) The textbook pays attention to the presentation of basic knowledge in the content, strictly follows the latest national and international standards, tries best to do less but better, with less class hours of teaching and more students self-study.

(2) The textbook highlights the scientific, systematic and practical nature of basic knowledge and theories, strengthens the cultivation of students' comprehensive ability, and enables students to master the ability to solve practical problems while laying a good theoretical foundation.

(3) The textbook adopts the latest national standards, and pays attention to the comparison between international standards and national standards, so that students can gradually adapt to the new requirements of the machinery manufacturing industry.

2. The compilation of the English content of the textbook

(1) The purpose of compiling the English-Chinese bilingual textbook is to make the students master the basic knowledge of interchangeability and measurement technology through the learning of English content, so as to lay a foundation for international communication in this field.

(2) In order to ensure the consistency of the basic content and the accuracy of

English content, the English part of the textbook does not adopt strict translation from Chinese content.

(3) The English content of the textbook, including diagrams, tables and formulas, is consistent with the Chinese content. Common vocabulary and primary scientific grammar are adopted as far as possible to ensure the readability of the English part of the textbook.

The Chinese part of this textbook is written by Chief Editor Wan Yipin, and the English part is written by Chief Editor Jia Jie. Professor Song Xudin was the Presiding Judge.

I would like to express my sincere gratitude to the leaders and colleagues of the Academic Affairs Office, the library and the School of Engineering Machinery of Chang'an University for their warm support and assistance in compiling this textbook.

Due to the limited level of the editors, the textbook will inevitably have omissions and impropery, especially in the English part of the content there are inevitably Chinese English expressions. We sincerely ask the readers for criticism and correction.

The Editor
In March 2021

前　言

"互换性与测量技术基础"是高等院校机械类、仪器仪表类和机电类各专业必修的主干技术基础课程之一，也是一门与机械工业发展紧密联系的基础学科。随着中国教育逐渐走向世界，急需编写具有中国特色且与国际接轨的教材。因此，在长安大学和人民交通出版社股份有限公司的支持下，编者开始尝试编写中英文双语互换性与测量技术基础教材。

基于课程教学课时少的特点，本教材侧重于对基本概念、基本理论、基本方法和基本技能的阐述，按照最新的国家标准和国际标准编写各章内容。

1. 关于本教材中文内容的编写说明

(1) 在内容上重视基础知识的讲述，严格遵循最新的国家标准和国际标准，尽量做到少而精，适合少课时的教学和学生自学。

(2) 突出基本知识和基本理论的科学性、系统性和实用性，加强对学生综合能力的培养，使学生打好理论基础的同时，掌握解决实际问题的能力。

(3) 教材采用现行国家标准，同时注重国际标准与国家标准的对比，使学生逐渐适应机械制造业的新要求。

2. 关于本教材英文内容的编写说明

(1) 编写英汉双语教材，其目的是让学生通过英文内容的学习，掌握互换性与测量技术基础的基本知识，为进行本专业领域内的国际交流奠定基础。

(2) 为保证基本内容的一致性和英文词义的准确性，教材英文部分的内容没有严格地按照汉语翻译。

(3) 教材的英文内容包括图、表、公式都与中文内容一致，尽可能采用常用词汇和初级科技语法，保证教材英文部分内容的可读性。

本教材中文部分由万一品编写，英文部分由贾洁编写。全书由万一品负责统稿，宋绪丁教授担任主审。

对给予本教材编写以热情支持和帮助的长安大学教务处、图书馆和工程机械学院的领导、同事表示诚挚的谢意。

由于编者水平有限，书中难免会有疏漏和不妥之处，特别是英文部分内容难免存在中式的英文表述，恳请广大读者批评指正。

编　者

2021 年 3 月

CONTENTS 目 录

Chapter 1　Introduction 绪论 ……………………………………………………………… 1
  1.1　Interchangeability 互换性 ……………………………………………………………… 1
  1.2　Standardization and Series of Preferred Values 标准化与优先数系 ………………… 3
  1.3　Purpose of This Course 本课程的学习任务 …………………………………………… 5
  Exercises 1 习题 1 …………………………………………………………………………… 6

Chapter 2　Size Tolerances and Fits 尺寸公差与配合 …………………………………… 7
  2.1　Basic Terms and Definitions 基本术语及定义 ………………………………………… 7
  2.2　Standards of Tolerances and Fits 公差与配合的国家标准 …………………………… 17
  2.3　Selection of Tolerances and Fits 公差与配合的选用 ………………………………… 39
  Exercises 2 习题 2 …………………………………………………………………………… 47

Chapter 3　Geometrical Tolerances 几何公差 …………………………………………… 49
  3.1　Basic Concepts 基本概念 ……………………………………………………………… 49
  3.2　Definition of Form Tolerances 形状公差 ……………………………………………… 56
  3.3　Profile Tolerances 轮廓度公差 ………………………………………………………… 60
  3.4　Orientation Tolerances 方向公差 ……………………………………………………… 62
  3.5　Location Tolerances 位置公差 ………………………………………………………… 65
  3.6　Run-out Tolerances 跳动公差 ………………………………………………………… 70
  3.7　Tolerance Principles 公差原则 ………………………………………………………… 74
  3.8　Selection of Geometrical Tolerance 几何公差的选用 ………………………………… 92
  Exercises 3 习题 3 …………………………………………………………………………… 101

Chapter 4　Surface Roughness 表面粗糙度 ……………………………………………… 102
  4.1　Basic Concept of Surface Roughness 表面粗糙度的基本概念 ……………………… 102
  4.2　Evaluation of Surface Roughness 表面粗糙度的评定 ………………………………… 105
  4.3　Parameter Value of Surface Roughness and Its Selection 表面粗糙度的参数值
       及其选用 ………………………………………………………………………………… 108
  4.4　Surface Roughness Symbol and Its Marking 表面粗糙度符号及其标注 …………… 112
  4.5　Measurement of Surface Roughness 表面粗糙度的检测 …………………………… 119
  Exercises 4 习题 4 …………………………………………………………………………… 121

Chapter 5　Fundation of Geometrical Quantity Measurement 几何量测量基础 ……… 122
  5.1　Introduction 概述 ……………………………………………………………………… 122
  5.2　Measurement Standard, Datum and Data 测量标准、基准和数据 ………………… 124
  5.3　Metrological Equipments and Measurement Methods 计量设备和测量方法 ……… 126

  5.4 Measurement Error and Data Processing 测量误差及数据处理 …………… 131

 Exercises 5 习题 5 ……………………………………………………………………… 134

Chapter 6 Tolerances and Fits for Typical Parts 典型零部件的公差与配合 ………… 135

  6.1 Cylindrical Gear 圆柱齿轮 …………………………………………………… 135

  6.2 Rolling Bearing 滚动轴承 ……………………………………………………… 138

  6.3 Key Connection 键连接 ………………………………………………………… 140

 Exercises 6 习题 6 ……………………………………………………………………… 141

References 参考文献 …………………………………………………………………… 142

Chapter 1　　Introduction 绪论

1.1　Interchangeability 互换性

1.1.1　Definition of Interchangeability 互换性的定义

The performance and the cost of production are always considered by engineers. However, usually the higher the performance is, the higher the cost is. And with the same performance, the higher the cost is, the lower the competitiveness in market is. How to ensure a production with high performance and low cost is the key problem for engineers. One answer is to design a mechanical component with interchangeability.

性能和成本是机械设计人员考虑的两大因素。但是,通常情况下产品性能和生产成本呈正比例关系。产品性能越优异,生产成本越高,反之亦然。处理产品性能与生产成本之间的矛盾,获得高性能、低成本的产品是设计人员必须考虑的,而互换性则是解决该矛盾的关键。

In daily life, there are a lot of devices with interchangeability. For example, when a component fails in a bicycle or a watch, we can buy a same specification and replace it, the bicycle and watch can be used again. In the mechanical engineering, there are more devices with interchangeability. The nuts, bolts and rolling bearing are all manufactured and assembled according to interchangeability principle.

日常生活中经常遇到关于互换性的实例。例如自行车、手表等零部件坏了,换上一个相同规格的新的零部件,能很好地满足使用要求。在机械工程领域,存在大量互换性的实例,例如螺母、螺栓和轴承。

Definition of interchangeability: interchangeability is the ability of a product, a process or a service to be used in place of another to fulfill the same requirements, without further machine or custom fitting. Interchangeability is an important principle of product design and manufacturing in machinery manufacturing industry.

互换性的定义:同一规格的一批零部件,任取其一,不需要任何加工和修配就能装在机器上,并满足相同的使用功能要求。互换性是机械制造行业中产品设计和制造的重要原则。

The interchangeability of one workpiece has two aspects: functional interchangeability and dimensional interchangeability. The mechanical parameters (such as strength, stiffness and hardness) and physical parameters (such as voltage, rotate speed and lighting) belong to functional interchangeability. The geometrical parameters (such as size, form and surface texture) belong to dimensional interchangeability. In this book, only the dimensional interchangeability is discussed.

机械制造中的互换性,通常包括性能互换和空间互换。常见的力、刚度和硬度等力学参数和电压、转速和亮度等物理参数属于性能互换。常见的尺寸、形状和表面纹理等几何参数属于空间互换。本教材只讨论空间互换。

Due to the error in machining, in order to make the parts interchangeable, it is necessary to control the error within the specified range. The range of allowable geometrical parameters of the part is called the tolerance, that is, the maximum allowable variation of the actual parameter value. The designer's task is to determine the correct tolerances and to clearly mark them on the drawings. The tolerance is the premise of interchangeability. In order to obtain the best technical and economic benefit on the basis of meeting functional requirements, the tolerance should be as large as possible.

由于加工时会产生误差,要使零件具有互换性,必须把零件的误差控制在规定的范围内。允许零件几何参数的变动范围称为公差,即实际参数值允许的最大变动量。设计者的任务就是正确地确定公差,并在图样上明确地标注出来。公差是实现互换性的前提,在满足功能要求的条件下,公差应该尽量规定得大一些,以获得最佳的技术经济效益。

1.1.2　Advantage and Classification 作用和分类

The role of interchangeability in machinery manufacturing:

(1) In designing: it can reduce the workload of drawing and calculation, shorten the design cycle, and is conducive to the application of computer-aided design.

(2) In manufacturing: a large number of standard machine elements can be made, then distributional manufactures can be adopted and automatic manufacturing lines can be applied, resulting in a lower cost of these standard machine elements.

(3) In repairing: a machine element can be replaced as quickly as possible when the machine element is interchangeable, to decrease the repairing time and cost, and to increase the efficiency and service time of the machine.

互换性在机械制造行业中的作用:

(1)在设计方面,零部件具有互换性可以减少绘图和计算工作量,缩短设计周期,并且有利于计算机辅助设计的应用。

(2)在制造方面,互换性有利于组织专业化生产,实现产品制造的自动化,从而降低生产成本。

(3)在维护方面,具有互换性的零部件损坏后可得到及时更换,从而减少了维修时间和费用,提高了机器的使用价值和寿命。

According to the interchange extent, interchangeability can be divided into two types: complete interchangeability and incomplete interchangeability. Complete interchangeability: in an assembly, no auxiliary options and repair are needed. Incomplete interchangeability: in an assembly, selection before assembly and adjustment during assembly are permitted.

互换性按照其互换程度可以分为完全互换和不完全互换两种。完全互换要求零部件在装配时不需要挑选和辅助加工。不完全互换允许在装配前对零件进行挑选、调整或辅助加工。

1.2 Standardization and Series of Preferred Values 标准化与优先数系

1.2.1 Standardization 标准化

In order to achieve interchangeability, a means must be adopted to maintain the necessary technical unification among the various scattered and local production departments and segments, so as to form a unified whole. Standardization is defined as: the development and implementation of concepts, doctrines, procedures and designs to achieve and maintain the required levels of compatibility, interchangeability or commonality in the operational, procedural, material, technical and administrative fields to attain interoperability. Standardization is one of the most important means of modern organization of production, the premise of professional collaboration of production, and an important part of scientific management.

为了实现互换性，必须采用一种手段，使各个分散的、局部的生产部门和生产环节之间保持必要的技术统一，以形成一个统一的整体。标准化的定义：对标准(概念、理论、程序和设计)的制定、贯彻、执行的一系列过程，以便在操作、程序、材料、技术和管理领域获得并保持所需的兼容性、互换性或通用性水平，以实现互操作性。标准化是组织现代化生产的一个重要手段，是实现专业化协调生产的必要前提，是科学管理的重要组成部分。

A standard approved by government departments, social organizations, etc., to promulgate unified specific provisions as the criterion and basis for common compliance, for example by corporation, regulatory body, military, etc. The standards adopted in mechanical industry mainly are international standards, national standards, local area standards, and industrial standards. All international standards start with ISO (International Standard Organization). Chinese standards start with GB (National Standards of the People's Republic of China). Machinery industry standards are with code JB. Enterprise standards start with QB.

标准是由政府部门、社会团体等机构批准的，以特定形式颁布的统一规定，作为公司、监管机构等共同遵守的准则和依据。标准通常分为国际标准、国家标准、地方/行业标准、企业标准等。国际标准冠以 ISO，国家标准冠以 GB，机械工业行业标准冠以 JB，企业标准冠以 QB。

Standardization is the important means to establish technology unity in modern production, and is also the technical basis for the implementation of the interchangeability of production.

标准化是现代生产中实现技术统一的重要手段，是实现互换性的前提，也是实现互换性的基础。

1.2.2 Series of Preferred Values 优先数系

It is an important part of standardization to simplify, coordinate and unify technical parameters. Many technical parameters are involved in product design and technical standard formulation. After selecting a value as the parameter index of a product, the value will be transferred according to certain rules. In production, the same parameters of the same product take different values from small to large, and form product series with different specifications to meet the various needs of users. Preferred numbers and preferred number series constitute a scientific numerical system, which provides an important basis for the numerical classification of product series.

工程上各种技术参数的简化、协调和统一是标准化的一项重要内容。在产品设计和技术标准制定时，涉及很多技术参数，选定一个数值作为某产品的参数指标后，这个数值就会按照一定的规律进行参数传递。生产中，为了满足用户各种各样的需求，同一种产品的同一参数从小到大取不同值，从而形成不同规格的产品系列。优先数和优先数系是一种科学的数值制度，它为产品系列的数值分级提供了重要基础。

In the 1870s, a French army officer, Colonel Charles Renard, tried to decrease the number of types of ropes for balloons and invented the series of preferred values. He derived a preferred values system using a geometric progression based on the number 10. The numbers created from the 5th (10th, 20th, 40th, 80th) root of 10, rounded off to one/two decimal places, created the R5 (R10, R20, R40, R80) series. R5, R10, R20 and R40 are the basic series, and R80 is the additional series. Their common ratio can be calculated according to the following equations:

19世纪70年代，法国人雷诺为减少系气球的绳索尺寸种类，按照等比数列发明了优先数系，数系每隔5项（10项、20项、40项、80项）数值增加为原来的10倍，得到R5（R10、R20、R40、R80）优先数系列。R5、R10、R20、R40为基本系列，R80为补充系列。5种优先数系的公比如下：

$$R5: q_5 = \sqrt[5]{10} \approx 1.5849, 取值1.60$$

$$R10: q_{10} = \sqrt[10]{10} \approx 1.2589, 取值1.25$$

$$R20: q_{20} = \sqrt[20]{10} \approx 1.1220, 取值1.12$$

$$R40: q_{40} = \sqrt[40]{10} \approx 1.0593, 取值1.06$$

$$R80: q_{80} = \sqrt[80]{10} \approx 1.0293, 取值1.03$$

The most common R5 series consists of these five rounded numbers:

最常见的R5系列5个数字为：

R5: 1.00　1.60　2.50　4.00　6.30

If our design constrains demand that the two screws in our gadget should be placed between 32mm and 55mm apart, we make it 40mm, because 4 is in the R5 series of preferred values.

如果一个产品设计要求的尺寸为32~55mm，通常选取该尺寸数值为40mm，因为4是R5系列的优先数。

The basic series of preferred values are shown in Tab. 1-1.

优先数系的基本系列见表1-1。

The basic series of preferred values Tab. 1-1
优先数系的基本系列 表1-1

R5	1.00						1.60					
R10	1.00			1.25			1.60			2.00		

Continued 续上表

R20	1.00		1.12		1.25		1.40		1.60		1.80		2.00		2.24	
R40	1.00	1.06	1.12	1.18	1.25	1.32	1.40	1.50	1.60	1.70	1.80	1.90	2.00	2.12	2.24	2.35
R5	2.50								4.00							
R10	2.50				3.15				4.00				5.00			
R20	2.50		2.80		3.15		3.55		4.00		4.50		5.00		5.60	
R40	2.50	2.65	2.80	3.00	3.15	3.35	3.55	3.75	4.00	4.25	4.50	4.75	5.00	5.30	5.60	6.00
R5	6.30								10.0							
R10	6.30				8.00				10.0							
R20	6.30		7.10		8.00		9.00		10.0							
R40	6.30	6.70	7.10	7.50	8.00	8.50	9.00	9.50	10.0							

The derived series: a priority number selected from every few items in the basic series to form a new series, if the basic series are not suitable the requirements, the derived series can be used. The derived method is as follows:

派生系列:从基本系列中每隔几项选取一个优先数,组成新的系列。如果基本系列的优先数满足不了工程实际要求,则需要用到派生系列。派生系列的生成方法为:

$$R\frac{i}{b} = \sqrt[b]{10^i} = 10^{i/b} \tag{1-1}$$

Where: b——the selected series value (for example $b=40$ for the R40 series);

i——the i-th element of this series (with the value of i between 0 and b).

式中:b——选取的基本系列(例如 $b=40$,即选取的 R40 优先数系);

i——间隔项数(i 的取值介于 0 和 b 之间)。

1.3 Purpose of This Course 本课程的学习任务

Interchangeability and Measurement Technology Foundation is the basic technique course for the students whose major is associated with mechanical. Engineers and workers who design or fabricate machines should have the knowledge introduced in this course. For the students who study in an engineer college, this course aims to train students to master the basic theory, knowledge and skills of product precision design and quality assurance, and lay the foundation for further application of national standards and control of product quality, the basic requirements are as follows:

互换性与测量技术基础是机械类专业学生的技术基础课。设计或制造机器的工程师和工人应具备本课程介绍的知识。对于在工科院校学习的学生,本课程旨在培养其掌握产品精度设计和质量保证的基本理论、知识和技能,为进一步应用国家标准和控制产品质量奠定基础,课程的基本要求如下:

(1) To understand clearly the concepts of interchangeability, standardization and measurement technology.

(2) To be familiar with the basic content and terms of every tolerance studied in this course.

(3) To choose the reasonable fit type according to the fit properties and to draw the size tolerance zone figure.

(4) During reading a detail drawing, to know exactly the precision requirements about size, form, orientation, location, run-out and surface texture. To be able to calculate the limits size, to describe the form, orientation, location requirements for every feature, to draw the dynamic tolerance zone figures under different tolerance principles.

(5) During drawing an engineer drawing (assembly drawing and detail drawing), to indicate the precision requirements on the drawing correctly according to the standards.

(1) 理解互换性、标准化和测量技术相关的基本概念。

(2) 熟悉并掌握课程中所讲各项公差的基本含义与基本概念。

(3) 能够根据配合性能选择合理的配合类型,并绘制尺寸公差带图。

(4) 掌握尺寸、形状、方向、位置、跳动和表面粗糙度的精度要求,能够计算尺寸的极限值,描述每个特征的形状、方向、位置要求,绘制不同公差原则下的动态公差图。

(5) 绘制工程图(装配图和零件图)时能够根据国家标准正确标注图样。

All kinds of tolerances have strict principles and regulations in the implementation of national standards, and have strong practicality in application. It should be noted that it is impossible to design the workpiece precision requirements only finish this course. To understand and apply all the knowledge properly need documented experience of several years in relative fields. So the students only study the fundamental knowledge about geometrical tolerance and measurement technology in this course.

各类公差在国家标准的贯彻上都有严格的原则性和法规性,而在应用时又具有很强的实践性。因此,学生通过本课程的学习,只能获得机械工程师所必需的互换性与测量技术基础知识与技能,而要牢固掌握和熟练应用,则需要毕业后积累一定的实际工作经验。

Exercises 1 习题 1

1-1 Please describe briefly the advantage and classification of interchangeability.

1-1 简要描述互换性的作用和分类。

1-2 Please describe briefly the conception of standardization.

1-2 简要描述标准化的概念。

1-3 The first number is 1, to write 5 numbers according to R5/3, R10, R10/2, R20/3.

1-3 优先数系起始值为1,试写出 R5/3、R10、R10/2 和 R20/3 系列的前 5 个优先数。

Chapter 2　Size Tolerances and Fits
尺寸公差与配合

2.1　Basic Terms and Definitions 基本术语及定义

2.1.1　Geometrical Features 几何要素

Geometrical feature: a general term applied to a physical portion of a part, such as a point, a line or a surface.

几何要素:构成零件几何特征的点、线、面统称为几何要素。

Size features: geometrical shape defined by a linear or angular size which is a size feature. The feature of size can be a cylinder, a sphere, two parallel opposite surfaces, a cone or a wedge.

尺寸要素:由一定大小的线性尺寸或角度尺寸确定的几何形状称为尺寸要素。尺寸要素可以是圆柱形、球形、两个平行对应面、圆锥形或楔形。

Integral feature: integral feature, also called the contour feature, refers to the surface or line on a surface, such as cylinder, plane, sphere, cone or contour line.

组成要素:组成要素也被称为轮廓要素。构成零件的面或面上的线称为组成要素,例如:圆柱面、平面、球面、圆锥面或轮廓线。

Derived feature: derived feature, also called the center feature, refers to center point, median line or median surface from one or more integral features.

导出要素:导出要素也被称为中心要素。由一个或多个组成要素得出对称中心所表示的点、线、面各要素统称为导出要素。

Nominal feature: nominal feature, also called the ideal feature, refers to the theoretically exact features as defined by engineer drawing or by other means that only have geometric meaning. All features indicated in the mechanical drawing are nominal features.

公称要素:公称要素也被称为理想要素,是指由工程图纸或其他方式定义的理论精确要素,在图样中只具有几何意义的要素。机械图样中所表示的要素均为公称要素。

Real feature: the feature that the part actually has is called the real feature. Due to the measurement error, the measured feature is not the actual condition of the real feature.

实际要素:零件实际存在的要素称为实际要素。由于存在测量误差,测得要素并非该实际要素的真实状况。

Measured feature: the feature that required the geometric tolerance given in the drawing or that needs to be studied and measured is called the measured feature.

被测要素:图样上给出了几何公差要求的要素或需要研究和测量的要素称为被测要素。

Datum feature: the feature used to determine the direction and the position of the measured feature is called datum feature. The nominal datum feature is referred to as the datum.

基准要素:用来确定被测要素方向和位置的要素称为基准要素,公称基准要素简称为基准。

2.1.2 Hole and Shaft 孔和轴

The fit tightness depends on the size of two features that the workpieces assembly. The two features of sizes are on two different workpieces. One size feature-containing is called hole and another feature-contained is called shaft. In a broad sense, hole and shaft can be either cylindrical or non-cylindrical.

在装配关系中,配合的紧密程度取决于在两个不同工件上的两个特征的尺寸。两个尺寸特征中,包容特征的是孔,被包容特征的是轴。广义上理解,孔和轴既可以是圆柱形的,也可以是非圆柱形的。

Hole: designates all cylindrical internal size features of a part, also including features of size which are not cylindrical.

孔通常指工件的圆柱形内尺寸要素,也包括非圆柱形的内尺寸要素。

Shaft: designates all cylindrical external size features of a part, also including features of size which are not cylindrical.

轴通常指工件的圆柱形外尺寸要素,也包括非圆柱形的外尺寸要素。

As shown in Fig. 2-1, the features indicated by D_1, D_2, D_3, D_4, D_5 and D_6 are holes, and the features indicated by d_1, d_2, d_3 and d_4 are shafts.

如图 2-1 所示,D_1、D_2、D_3、D_4、D_5、D_6 为孔的尺寸,d_1、d_2、d_3、d_4 为轴的尺寸。

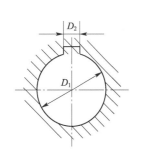
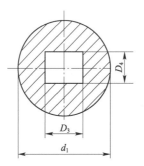
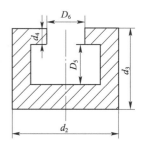

Fig. 2-1 Diagram of Hole and Shaft Definition

图 2-1 孔和轴

2.1.3 Terms and Definitions about Size 有关尺寸的术语与定义

Size is a number expressing, in a particular unit, the numerical value of a linear size, such as diameters, radius, length, width, thickness, etc. Usually, the unit is mm, and in this textbook, when the unit is millimeter, it is not written out on the technical drawings.

尺寸是指用特定单位表示线性尺寸值的数值,例如直径、半径、长度、宽度、高度等。通常尺寸的单位是毫米(mm),以 mm 为单位时,可以只写数值不写单位。

Nominal size, also called basic size, refers to the size of a feature of perfect form as defined by the drawing specification. Nominal size is the theoretical size from which the hole/shaft limits are derived. Nominal size is given during the process of designing according to the experience. Capital letter D is used to present hole basic size, and lower case letter d is used to present shaft nominal size.

公称尺寸也被称为基本尺寸。公称尺寸是指由图样规范确定的理想要素尺寸。通过公称尺寸可以用极限偏差计算出极限尺寸。公称尺寸是设计者通过计算或根据经验确定的尺寸。孔的公称尺寸用大写字母 D 表示，轴的公称尺寸用小写字母 d 表示。

Actual size refers to the size of associated integral feature, and the actual size is obtained by measurement. Usually, the actual sizes at different places of a feature of size are different due to the measurement error. The code D_a and d_a represent the actual size of holes and shafts respectively.

实际尺寸是指关联组成要素的尺寸，是通过测量获得的局部尺寸。由于误差的影响，在零件同一表面不同位置测得的实际尺寸是不相同的。孔和轴的实际尺寸分别用 D_a 和 d_a 表示。

Limits size refer to the two extreme permissible size of a feature, between which the actual size should be, the limits size being included. Include upper limit size and lower limit size.

极限尺寸是指尺寸要素允许的尺寸的两个极端值，实际尺寸应介于极限尺寸之间，包括上极限尺寸和下极限尺寸。

(1) Maximum limit size: the largest permissible size of a size feature. D_{max} and d_{max} represent the maximum limit size for holes and shafts.

(2) Minimum limit size: the smallest permissible size of a size feature. D_{min} and d_{min} represent the minimum limit size for holes and shafts.

(1) 上极限尺寸：尺寸要素允许的最大尺寸。孔和轴的上极限尺寸分别用 D_{max} 和 d_{max} 表示。

(2) 下极限尺寸：尺寸要素允许的最小尺寸。孔和轴的下极限尺寸分别用 D_{min} 和 d_{min} 表示。

Normally, any actual size usually should not exceed the maximum limit size and the minimum limit size, i.e., $D_{min} \leq D_a \leq D_{max}$, $d_{min} \leq d_a \leq d_{max}$, otherwise, the workpiece is unqualified. The limit size is used to control the actual size.

一般情况下，工件任意一处的实际尺寸是不能超过上下极限尺寸所组成的范围，若超出极限尺寸范围，对应的工件尺寸是不合格的。极限尺寸被用来控制实际尺寸。

2.1.4 Deviation and Tolerance 偏差与公差

2.1.4.1 Deviation 偏差

Deviation: value minus its datum value. For size deviations, the value is the actual size or limt size and the datum value is the nominal size. Then the size deviation can be defined as the algebraic difference of a size minus the nominal size. Size deviations include actual deviation and limit deviation.

某一尺寸减其公称尺寸所得的代数差,称为偏差。这里的"某一尺寸"是指提取实际组成要素的局部尺寸或极限尺寸。根据"某一尺寸"的不同,偏差可分为实际偏差和极限偏差。

Actual deviation refers to the the algebraic different value between actual local size and the nominal size. E_a and e_a represent actual deviation of holes and shafts.

实际偏差:实际(组成)要素的局部尺寸减其公称尺寸所得的代数差。孔和轴的实际偏差分别用符号 E_a 和 e_a 表示。用公式表示为:

$$E_a = D_a - D \tag{2-1}$$

$$e_a = d_a - d \tag{2-2}$$

Limit deviation refers to the algebraic difference between limits size and nominal size, which includes upper limit deviation and lower limit deviation. Upper limit deviation is the algebraic difference between upper limits size and nominal size, and lower limit deviation is the algebraic difference between lower limits size and nominal size. ES and EI represent upper limit deviation and lower deviation of holes, and es and ei represent upper limit deviation and lower deviation of shafts, the equations are as follows:

极限偏差指上极限偏差和下极限偏差,用于限制实际组成要素偏差。上极限尺寸减其公称尺寸所得的代数差称为上极限偏差。孔和轴的上极限偏差分别用符号 ES 和 es 表示。下极限尺寸减其公称尺寸所得的代数差称为下极限偏差。孔和轴的下极限偏差分别用符号 EI 和 ei 表示。孔、轴上下极限偏差分别用以下公式表示为:

$$\text{ES} = D_{\max} - D \tag{2-3}$$

$$\text{EI} = D_{\min} - D \tag{2-4}$$

$$\text{es} = d_{\max} - d \tag{2-5}$$

$$\text{ei} = d_{\min} - d \tag{2-6}$$

2.1.4.2　Size Tolerance 尺寸公差

Size tolerance is the difference between the upper limit size and the lower limit size, and also is the difference between the upper limit deviation and lower limit deviation. Size tolerance is an absolute quantity without sign. T_D and T_d represent size tolerance of holes and shafts, the Equations are as follows:

尺寸公差简称公差,指上极限尺寸与下极限尺寸之差,或上极限偏差与下极限偏差之差。它是允许的尺寸变动量。孔和轴的公差分别用 T_D 和 T_d 表示,其计算公式为:

$$T_D = |D_{\max} - D_{\min}| = |\text{ES} - \text{EI}| \tag{2-7}$$

$$T_d = |d_{\max} - d_{\min}| = |\text{es} - \text{ei}| \tag{2-8}$$

The size tolerance is the variation value, so it is an absolute value without sign. Size tolerance value cannot be zero.

尺寸公差是一个没有符号的绝对值,且不能为0。

The relationship among nominal size, limit size, limit deviation and tolerance is shown in Fig. 2-2.

公称尺寸、极限尺寸、极限偏差和公差之间的相互关系如图2-2所示。

Chapter 2 Size Tolerances and Fits 尺寸公差与配合

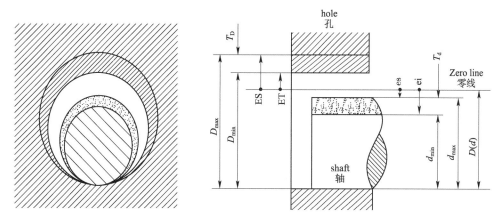

Fig. 2-2 Relationship among nominal size, limit size, limit deviation and tolerance
图 2-2 公称尺寸、极限尺寸、极限偏差和公差之间的相互关系

2.1.4.3 Illustration of Limits and Fits 公差带图

To simplify, the body of holes and shafts is not drawn, and only the tolerance zone is drawn enlarged to analyze the relationship of the holes and shafts, as shown in Fig. 2-3. This figure is named illustration of limits and fits, and also named size tolerance zone. The illustration of limits and fits includes zero line and size tolerance zone.

直观表示出公称尺寸、极限偏差、公差以及孔与轴配合关系的图解,简称公差带图,如图 2-3 所示。公差带图包含零线和公差带。

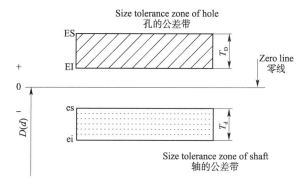

Fig. 2-3 Illustration of Limits and Fits
图 2-3 公差带图

Zero line: in an illustration of limits and fits, the straight line, representing the nominal size, to which the deviations and tolerances are referred. The zero line is drawn horizontally, with positive deviations shown above and negative deviations below.

零线:在公差带图中,表示公称尺寸的一条直线称为零线,以其为基准确定偏差和公差。正偏差位于零线的上方,负偏差位于零线的下方。

Size tolerance zone: in an illustration of limits and fits, the zone contained between two lines representing the upper and lower deviations, defined by the magnitude the size tolerance zone and its position of the size tolerance zone. The magnitude of the tolerance zone is the width of size tolerance zone in the direction perpendicular to the zero line, which is determined by standard tolerance. The position of tolerance zone is relative to the zero line in the direction of perpendicular to the zero line, which is determined by the fundamental deviation.

公差带:在公差带图中,由代表上、下极限偏差或上、下极限尺寸的两条直线所限定的一个区域,称为公差带。公差带有两个基本参数,即公差带大小与公差带位置。公差带大小由标准公差确定,公差带位置由基本偏差确定。

Standard tolerance refers to any tolerance belongs to GB standard system of limits and fits.

标准公差:国标表格中所列的用以确定公差带大小的任一公差值都是标准公差。

Fundamental deviation refers to the deviation which defines the position of the tolerance zone in relation to the zero line in the GB system of limits and fits. The fundamental deviation may be either the upper limit deviation or the lower limit deviation. Usually the fundamental deviation is the one nearer to the zero line.

基本偏差:国标极限配合制中,确定公差带相对零线位置的那个极限偏差。它可以是上极限偏差或下极限偏差,一般为靠近零线的那个偏差。

Example 2-1 Given: the nominal size of hole and shaft is 25mm, the limits size for hole is $D_{max} = 25.021$mm and $D_{min} = 25.000$mm, the limits size for shaft is $d_{max} = 24.980$mm and $d_{min} = 24.967$mm, answer the following questions: (1) to calculate the limit deviations and size tolerance; (2) to draw the illustration of size tolerance zone.

例2-1 已知孔、轴公称尺寸为25mm, $D_{max} = 25.021$mm, $D_{min} = 25.000$mm, $d_{max} = 24.980$mm, $d_{min} = 24.967$mm。(1)求孔与轴的极限偏差和公差;(2)画出尺寸公差带图。

Solution:

解:

(1) According to Equation (2-3) to Equation (2-6), the limit deviation can be achieved:

(1)根据式(2-3)~式(2-6)可得:

Upper deviation of hole:

孔的上极限偏差:

$$ES = D_{max} - D = 25.021 - 25 = +0.021(\text{mm})$$

Lower deviation of hole:

孔的下极限偏差:

$$EI = D_{min} - D = 25 - 25 = 0(\text{mm})$$

Upper deviation of shaft:

轴的上极限偏差:

$$es = d_{max} - d = 24.980 - 25 = -0.020(\text{mm})$$

Lower deviation of shaft:

轴的下极限偏差:

$$ei = d_{min} - d = 24.967 - 25 = -0.033(\text{mm})$$

(2) According to Equation (2-7) to Equation (2-8), the size tolerance of hole and shaft can be achieved:

(2)根据式(2-7)~式(2-8)可得:

The size tolerance of hole:

孔的公差：

$T_D = |D_{max} - D_{min}| = |ES - EI| = 0.021(mm)$

The size tolerance of shaft:

轴的公差：

$T_d = |d_{max} - d_{min}| = |es - ei| = 0.013(mm)$

The illustration of size tolerance zone is shown in Fig. 2-4.

孔、轴尺寸公差带图如图 2-4 所示。

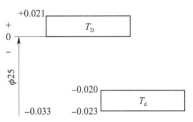

Fig. 2-4　Illustration of size tolerance zone

图 2-4　尺寸公差带图

2.1.5　Terms and Definition about Fits 有关配合的术语和定义

2.1.5.1　Fits 配合

Fit refers to the relationship between size tolerance zones of holes and shafts assembled together, of which nominal sizes are same. According to the definition of fits, there are two fundamental conditions: one is that the nominal size should be same, i. e. $D = d$; another is that the holes and shafts have the relationship of containing and contained, i. e. they are assembled together. At the same time, the fits are the assembly relationships of a batch of holes and shafts, rather than one certain hole assemble with one certain shaft, so it is exact to describe fits using the relationship of size tolerance zone.

配合是指公称尺寸相同、相互结合的孔与轴公差带之间的关系。形成配合要有两个基本条件：一是孔和轴的公称尺寸必须相同，二是具有包容或被包容的特性，即孔与轴的结合。配合是指一批孔、轴的装配关系，而不是指单个孔和单个轴的相配，所以用公差带相互位置关系来反映配合比较确切。

2.1.5.2　Clearance and Interference 间隙和过盈

Clearance is the positive difference between the sizes of the hole and the shaft. Before assembly, when the size of the shaft is smaller than the size of the hole. X is used to represent clearance value. Interference is the negative difference between the sizes of the hole and the shaft. Before assembly, when the size of the shaft is larger than the size of the hole, Y is used to represent clearance value. It should be pointed out that, to calculate clearance or interference, always the size of hole minus the size of shaft, if the result is positive then the clearance is achieved, and if the result is negative then the interference is achieved. The result is positive only means the size of the hole is larger than the shaft, and is negative only means the size of hole is smaller than the size of shaft. In other words, interference is negative clearance, and clearance is negative interference. The clearance value determines the relative movement of two fit parts, and the interference value determines the tightness of the connection.

孔的尺寸减去相配合的轴的尺寸所得的代数差，此值为正时是间隙，用 X 来表示，为负时是过盈，用 Y 来表示。过盈就是负的间隙，间隙也就是负的过盈。间隙大小决定两相配合零件相对运动的活动程度，过盈大小则决定两相配合零件连接的牢固程度。

2.1.5.3　Fits Types 配合种类

According to the position relationship of the holes and shafts, there are three types of fits,

i. e. clearance fits, interference fits and transition fits, as shown in Fig. 2-5.

配合按其出现间隙、过盈的不同,可以分为间隙配合、过盈配合和过渡配合3个类别,如图2-5所示。

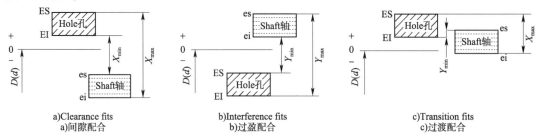

Fig. 2-5 Schematic Representation of Fits Types

图2-5 配合类型示意图

(1) Clearance fit.

(1) 间隙配合。

A fit that always provides a clearance between the hole and the shaft when assembled, i. e. the minimum size of the hole is either greater than or, in the extreme case, equal to the maximum size of the shaft. The size tolerance zone of hole is totally above the size tolerance zone of shaft, as shown in Fig. 2-5a).

间隙配合是指具有间隙(包括最小间隙等于0)的配合。间隙配合中孔的公差带在轴的公差带之上,如图2-5a)所示。

For a clearance fit, the limits of clearance are concerned. The limits of clearance is the general term of the maximum clearance and the minimum clearance are the limits of the allowed clearance. The clearance varied from minimum clearance to maximum clearance. Sometimes, the average clearance is used to describe the property of the clearance fit.

在间隙配合中,孔和轴有两个极限尺寸,因而间隙也有最大间隙和最小间隙。间隙配合的性质用最大间隙代数量和最小间隙代数量表示,有时也用平均间隙来表示。

①Maximum clearance: in a clearance fit, the positive difference between the maximum limit size of the hole and the minimum limit size of the shaft, designated by X_{max}.

②Minimum clearance: in a clearance fit, the positive difference between the minimum limit size of the hole and the maximum limit size of the shaft, designated by X_{min}.

③Average clearance: the average value of maximum clearance and minimum clearance, designated by X_{av}.

①孔的上极限尺寸减轴的下极限尺寸之差,称为最大间隙 X_{max}。

②孔的下极限尺寸减轴的上极限尺寸之差,称为最小间隙 X_{min}。

③最大间隙和最小间隙的平均值称为平均间隙 X_{av}。

$$X_{max} = D_{max} - d_{min} = (D + ES) - (d + ei) = ES - ei \tag{2-9}$$

$$X_{min} = D_{min} - d_{max} = (D + EI) - (d + es) = EI - es \tag{2-10}$$

$$X_{av} = (X_{max} + X_{min})/2 \tag{2-11}$$

A plus sign must be marked before the calculated value of clearance fit.

间隙值的前面必须标注正号。

(2) Interference fit.

(2)过盈配合。

A fit which everywhere provides an interference between the hole and the shaft when assembled, i. e. the maximum size of the hole is either smaller than or, in the extreme case, equal to the minimum size of the shaft. The size tolerance zone of shaft is totally above the size tolerance zone of hole, as shown in Fig. 2-5b).

过盈配合是指具有过盈(包括最小过盈等于0)的配合。过盈配合中孔的公差带在轴的公差带之下,如图2-5b)所示。

For an interference fit, the limits of interference are concerned. Limits of interference is the general term of the maximum interference and the minimum interference, they are the limits of the allowed interference. The interference varied from minimum interference to maximum interference. Sometimes, the average interference is used to describe the property of the interference fit.

在过盈配合中,由于孔和轴各有两个极限尺寸,因而过盈也有最大过盈和最小过盈。过盈配合的性质用最大过盈代数量和最小过盈代数量表示,有时也用平均过盈来表示。

①Maximum interference: in an interference fit, the negative difference between the minimum limit size of the hole and the maximum limit size of the shaft, designated by Y_{max}.

②Minimum interference: in an interference fit, the negative difference between the maximum limit size of the hole and the minimum limit size of the shaft, designated by Y_{min}.

③Average interference: the average value of maximum interference and minimum interference, designated by Y_{av}.

①孔的下极限尺寸减轴的上极限尺寸之差,称为最大过盈Y_{max}。

②孔的上极限尺寸减轴的下极限尺寸之差,称为最小过盈Y_{min}。

③最大过盈和最小过盈的平均值称为平均过盈Y_{av}。

$$Y_{max} = D_{min} - d_{max} = (D + EI) - (d + es) = EI - es \quad (2\text{-}12)$$

$$Y_{min} = D_{max} - d_{min} = (D + ES) - (d + ei) = ES - ei \quad (2\text{-}13)$$

$$Y_{av} = (Y_{max} + Y_{min})/2 \quad (2\text{-}14)$$

A minus sign must be marked before the calculated value of interference fit.

过盈值的前面必须标注负号。

(3) Transition fit.

(3)过渡配合。

A fit which may provide either clearance or interference between the hole and the shaft when assembled, depending on the actual sizes of the hole and the shaft, i. e. the tolerance zones of the hole and the shaft overlap completely or in part, as shown in Fig. 2-5c).

过渡配合是指可能具有间隙或过盈的配合。过渡配合中孔的公差带和轴的公差带相互交叠,如图2-5c)所示。

For a transition fit, the limit interference and clearance are concerned. Limit clearance and interference are the general terms of the maximum clearance and the maximum interference, and they are the limits of the allowed clearance and interference. The clearance and interference varied from maximum clearance to maximum interference. The average value of the limit clearance and interference may be plus or minus.

在过渡配合中,其配合的性质用最大间隙代数量和最大过盈代数量来表示。最大间隙和最大过盈的平均值是间隙还是过盈取决于平均值的符号,为正时是平均间隙,为负时是平均过盈。

2.1.5.4 Fit Tolerances 配合公差

Fit tolerance is the variation of a fit, i.e. the absolute value of the difference between limit clearance(s) and/or limit interference(s), as Equation(2-15) to Equation(2-17). T_f also equals to the arithmetic sum of the tolerance of the two features composing the fit, as Equation(2-15).

组成配合的孔、轴公差之和,是允许间隙或过盈的变动量,用 T_f 表示。对于间隙配合,配合公差等于最大间隙与最小间隙之差的绝对值;对于过盈配合,配合公差等于最小过盈与最大过盈之差的绝对值;对于过渡配合,配合公差等于最大间隙与最大过盈之差的绝对值。计算公式如式(2-15)~式(2-17)所示。

$$\text{Clearance fit 间隙配合}: T_f = |X_{max} - X_{min}| \qquad (2\text{-}15)$$

$$\text{Interference fit 过盈配合}: T_f = |Y_{min} - Y_{max}| \qquad (2\text{-}16)$$

$$\text{Transition fit 过渡配合}: T_f = |X_{max} - Y_{max}| \qquad (2\text{-}17)$$

X_{max}, X_{min}, Y_{max} and Y_{min} in Equation (2-15) to Equation(2-17) are substituted by upper deviation and lower deviation, then

用极限偏差表示上述式中的 X_{max}、X_{min}、Y_{max} 和 Y_{min} 可以得到

$$T_f = |X_{max}(Y_{min}) - X_{min}(Y_{max})| = |(ES - ei) - (EI - es)|$$
$$= |(ES - EI) + (es - ei)| = T_D + T_d$$

i.e. the equations of fit tolerance for all three fit types are the same:

即3种配合类型的配合公差公式相同:

$$T_f = T_D + T_d \qquad (2\text{-}18)$$

Example 2-2 Given: a hole which size and deviation is $\phi 30^{+0.033}_{0}$ fit with following shaft, to calculate the limit clearance(s) and/or interference(s), the average clearance or interference, and the fit tolerance.

(1) $\phi 30^{-0.020}_{-0.041}$;

(2) $\phi 30^{+0.069}_{+0.048}$;

(3) $\phi 30^{+0.013}_{-0.008}$;

(4) Then to draw the illustration of limits and fits.

例 2-2

(1) 计算孔 $\phi 30^{+0.033}_{0}$ 与轴 $\phi 30^{-0.020}_{-0.041}$ 配合的 X_{max}、X_{min}、X_{av} 和 T_f;

(2) 计算孔 $\phi 30^{+0.033}_{0}$ 与轴 $\phi 30^{+0.069}_{+0.048}$ 配合的 Y_{max}、Y_{min}、Y_{av} 和 T_f;

(3) 计算孔 $\phi 30^{+0.033}_{0}$ 与轴 $\phi 30^{+0.013}_{-0.008}$ 配合的 X_{max}、Y_{max}、X_{av} 和 T_f;

(4) 画出上述孔与轴配合的公差带图。

Solution:

解:

(1) According to Equation (2-9) to Equation (2-11) and Equation (2-18):

(1) 由式(2-9)到式(2-11)和式(2-18)可知：
$$X_{max} = ES - ei = (+0.033) - (-0.041) = +0.074$$
$$X_{min} = EI - es = 0 - (-0.020) = +0.020$$
$$X_{av} = (X_{max} + X_{min})/2 = [(+0.074) + (+0.020)]/2 = +0.047$$
$$T_f = T_D + T_d = |(ES - EI) + (es - ei)| = 0.054$$

(2) According to Equation (2-12) to Equation (2-14) and Equation (2-18):
(2) 由式(2-12)到式(2-14)和式(2-18)可知：
$$Y_{max} = EI - es = 0 - (+0.069) = -0.069$$
$$Y_{min} = ES - ei = (+0.033) - (+0.048) = -0.015$$
$$Y_{av} = (Y_{max} + Y_{min})/2 = [(-0.069) + (-0.015)]/2 = -0.042$$
$$T_f = T_D + T_d = |(ES - EI) + (es - ei)| = 0.054$$

(3) According to Equation (2-9), Equation (2-12), Equation (2-11) and Equation (2-18):
(3) 由式(2-9)、式(2-12)、式(2-11)和式(2-18)可知：
$$X_{max} = ES - ei = (+0.033) - (-0.008) = +0.041$$
$$Y_{max} = EI - es = 0 - (+0.013) = -0.013$$
$$X_{av} = (X_{max} + X_{min})/2 = [(+0.041) + (-0.013)]/2 = +0.014$$
$$T_f = T_D + T_d = |(ES - EI) + (es - ei)| = 0.054$$

(4) The illustration is shown in Fig. 2-6.
(4) 孔与轴配合的公差带如图2-6所示。

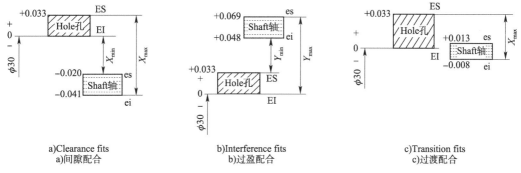

Fig. 2-6 Illustration of Limits and Fits
图 2-6 公差带图

2.2 Standards of Tolerances and Fits 公差与配合的国家标准

The combination of the hole and the shaft is used widely in mechanical devices. According to the functional requirements, there are three fit types when a hole and a shaft are assembled together: clearance fit used for the connection that is able to move relatively, interference fit used for permanent connection, and transition fit used for the connection that is able to locate and remove. To meet these three functional requirements, *Geometrical product specifications (GPS)—ISO code system for tolerances on linear sizes—Part 1: Basis of tolerances, deviations and fits* (GB/T 1800.1—2020) specifies fit system, standard tolerance series and fundamental deviation series.

孔与轴的组合形式在机械设备中应用广泛。根据功能要求,孔与轴的装配有三种方式,间隙配合用于相对运动的连接,过盈配合用于固定的连接,过渡配合用于定心可拆卸的连接。《产品几何技术规范(GPS) 线性尺寸公差 ISO 代号体系 第1部分:公差、偏差和配合的基础》(GB/T 1800.1—2020)规定了配合制、标准公差系列和基本偏差系列。

2.2.1 Fit System 配合制

The fit system refers to a system in which the holes and the shafts of the same limit system are formed into a fit. There are two basis fit systems in GB system: hole-basis fit system and shaft-basis fit system.

配合制是指同一极限制的孔和轴组成配合的一种制度。国标中规定了两种等效的配合制:基孔制配合和基轴制配合。

Hole-basis fit system is a fit system in which the required different fit properties are obtained by combining shafts of various fundamental deviations with the hole of a fixed fundamental deviation, shown in Fig. 2-7a). The hole in the Hole-basis fit system is a reference hole, and its code is "H"; the lower limit size of this hole is identical to the nominal size and the fundamental deviation is lower deviation, i. e. EI = 0.

基孔制配合是指基本偏差为一定的孔的公差带,与不同基本偏差的轴的公差带形成各种配合的一种制度。基孔制配合中的孔为基准孔,其代号为 H。基孔制配合中孔的下极限尺寸与公称尺寸相等,孔的下极限偏差(EI)为 0,如图 2-7a)所示。

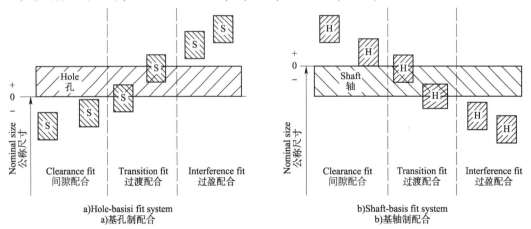

Fig. 2-7 Schematic of Fits System

图 2-7 配合制示意图

Shaft-basis fit system is a fit system in which the required different fit properties are obtained by combining holes of various fundamental deviations with the shaft of a fixed fundamental deviation, shown in Fig. 2-7b). The shaft in the Shaft-basis fit system is a reference shaft, and its code is "h"; the upper limit size of this shaft is identical to the nominal size and the fundamental deviation is upper deviation, i. e. es = 0.

基轴制配合是指基本偏差为一定的轴的公差带,与不同基本偏差的孔的公差带形成各种配合的一种制度。基轴制配合中的轴为基准轴,其代号为 h。基轴制配合中轴的上极限尺寸与公称尺寸相等,轴的上极限偏差(es)为 0,如图 2-7b)所示。

2.2.2 Standard Tolerance Series 标准公差系列

Standard tolerance is any tolerance value prescribed by the national standard to determine the size of the tolerance zone. Standard tolerance values reflect the accuracy of the workpieces, and they also reflect the machining difficulty and costs. Therefore, different tolerance grades are provided to meet different precision requirements. Moreover, at the same tolerance grade, the machining difficulty should be the same at different sizes. Then the tolerance values are relevant to two factors: tolerance grades and nominal sizes.

标准公差是国家标准规定的用以确定公差带大小的任一公差值。标准公差值反映了工件的加工精度,同时也反映了加工难度和加工成本。因此,国标提供了不同的公差等级来满足不同的精度要求。在相同公差等级下,不同尺寸的加工难易程度是相同的。标准公差值大小与公差等级和公称尺寸两个因素相关。

2.2.2.1 Standard Tolerance Factors 标准公差因子

In a standard system of limits and fits, tolerances series are considered as corresponding to the same level of accuracy for all nominal sizes. It is the name given to one standard series of tolerances calculated according to certain law in terms of the nominal size. Thus, it is a function of nominal size and is common to the two formulae defining the different grades of tolerance. Standard tolerance factor i (I) is the basic unit for calculating standard tolerance value, and also the basis for formulating standard tolerance series. According to the production practice, special scientific test and statistical analysis, it is shown that there is a certain relationship between the standard tolerance factor and the nominal sizes.

在公差与配合标准系统中,公差系列被看作是所有公称尺寸所具有的相同精度水平。它是公差的一个标准系列名称,该公差系列是基于公称尺寸并根据一定规律计算得到的。因此,它是公称尺寸的函数,也是定义不同公差等级的两个公式。标准公差因子$i(I)$是计算标准公差值的基本单位,也是制定标准公差系列的基础。生产实践以及专门的科学试验和统计分析表明,标准公差因子与零件尺寸之间有一定的关系。

The standard tolerance factor, i, in micrometers, is calculated from the Equation (2-19) for the nominal size up to 500mm.

公称尺寸值小于或等于500mm时,标准公差因子i(单位:μm)计算式如式(2-19)所示。

$$i = 0.45\sqrt[3]{D} + 0.001D \qquad (2\text{-}19)$$

Where: D——Geometric mean of the first and last sizes of the size segment in which the nominal size is located (mm);

i——Standard tolerance factor (μm).

式中:D——公称尺寸所在尺寸段的首尾尺寸的几何平均值(mm);

i——标准公差因子(μm)。

The standard tolerance factor, I, in micrometers, is calculated from the following formula for the basic size from 500mm up to 3150mm.

公称尺寸值为500~3150mm时,标准公差因子I(单位:μm)计算式如式(2-20)所示。

$$I = 0.004D + 2.1 \tag{2-20}$$

Where: I——Standard tolerance factor (μm).

式中:I——标准公差因子(μm)。

2.2.2.2 Standard Tolerance Grades 公差等级

The standard tolerance stipulated by the national standard is determined by the product value of the tolerance grade coefficient and the standard tolerance factor. When the nominal size is fixed, the coefficient of tolerance grade is the only parameter to determine the size of standard tolerance.

国家标准规定的标准公差是由公差等级系数和标准公差因子的乘积值决定的。在公称尺寸一定的情况下,公差等级系数是决定标准公差大小的唯一参数。

The GB system of limits and fits provides 20 standard tolerance grades in the size range from 0 to 500 mm (including 500mm) designated IT01, IT0, IT1, ..., IT18, and 18 standard tolerance grades in the size range from 500mm up to 3150mm (including 3150mm) designated IT1, IT2, ..., IT18. Arabic numerals are added after IT, which represents standard tolerances, and Arabic numerals are standard tolerance grades. Among them, IT01 possesses the highest precision and the least tolerance, while IT18 possesses the lowest precision and the largest tolerance. The lower the grade is, the larger the tolerance will be. The values of standard tolerance are calculated from formulas given in Tab. 2-1.

根据公差等级系数的不同,国家标准规定 0~500mm 尺寸段标准公差分为 20 个等级,即 IT01、IT0、IT1、IT2、……、IT18;500~3150mm 尺寸段标准公差分为 18 个等级,即 IT1、IT2、……、IT18。IT 后加阿拉伯数字,IT 表示标准公差,阿拉伯数字为标准公差等级。从 IT01 到 IT18,公差等级依次降低,公差依次增大。不同公差等级下标准公差值的计算公式见表 2-1。

Formulae for Standard Tolerance   Tab. 2-1
各级标准公差计算公式   表 2-1

Tolerance grades 公差等级	Nominal size 公称尺寸(mm)		Tolerance grades 公差等级	Nominal size 公称尺寸(mm)	
	>0~500	>500~3150		>0~500	>500~3150
IT01	$0.3 + 0.008D$	No these grades 不存在	IT7	$16i$	$16I$
IT0	$0.5 + 0.012D$		IT8	$25i$	$25I$
IT1	$0.8 + 0.020D$	$2I$	IT9	$40i$	$40I$
IT2	(IT1)$\left(\frac{\text{IT5}}{\text{IT1}}\right)^{1/4}$	$2.7I$	IT10	$64i$	$64I$
IT3	(IT1)$\left(\frac{\text{IT5}}{\text{IT1}}\right)^{1/2}$	$3.7I$	IT11	$100i$	$100I$
IT4	(IT1)$\left(\frac{\text{IT5}}{\text{IT1}}\right)^{3/4}$	$5I$	IT12	$160i$	$160I$
IT5	$7i$	$7I$	IT13	$250i$	$250I$
IT6	$10i$	$10I$	IT14	$400i$	$400I$

Continued 续上表

Tolerance grades 公差等级	Nominal size 公称尺寸(mm)		Tolerance grades 公差等级	Nominal size 公称尺寸(mm)	
	>0~500	>500~3150		>0~500	>500~3150
IT15	640i	640I	IT17	1600i	1600I
IT16	1000i	1000I	IT18	2500i	2500I

Note: the results are in micrometers and the nominal sizes are in millimeters.
注:公称尺寸的单位为 mm,标准公差值计算结果单位为 μm。

2.2.2.3 Segments of Nominal size 尺寸分段

According to the calculation formula of standard tolerance, for each tolerance grade, each nominal size can get a corresponding tolerance value. However, there are many nominal sizes in production, so the tolerance table compiled in this way will be extremely huge, which will bring trouble to the actual production and is not conducive to the standardization and serialization of tolerance values. In order to reduce the number of tolerances, simplify the tolerance table and facilitate the application, the national standard has segmented the nominal size.

按照标准公差的计算公式,对于每个公差等级,每个公称尺寸都可计算出一个相应的公差值。但在生产中公称尺寸很多,这样编制的公差表格将会极为庞大,给生产实际带来麻烦,也不利于公差值的标准化和系列化。为了减少公差数目,简化公差表格,便于应用,国标对公称尺寸进行了分段。

For convenience, the standard tolerances and fundamental deviations are not calculated individually for each separate nominal size, but for segments of the nominal size as given in Tab. 2-2. The values of the standard tolerances and fundamental deviations for each nominal size segment are calculated from the geometrical mean (D) of the extreme sizes ($>D_n \sim D_{n-1}$) of that segment, as follows: $D = (D_n \cdot D_{n-1})^{1/2}$. For the first nominal size segment ($\leq 3mm$), the geometrical mean, D, according to convention, is taken between size 1mm and 3mm, therefore $D = 1.732mm$. The values of standard tolerance for some nominal sizes are given in Tab. 2-2.

在标准公差和基本偏差的计算公式中,公称尺寸一律以所属分段尺寸的首尾两项的几何平均值进行计算,尺寸分段($>D_n \sim D_{n-1}$)的首尾两项的几何平均值 $D = (D_n \cdot D_{n-1})^{1/2}$,但对于 ≤3mm 的尺寸段,$D = (1 \times 3)^{1/2} = 1.732mm$。由标准公差数值构成的表格为标准公差数值表,见表 2-2。

Example 2-3 Given that the nominal size is 20mm, find the tolerance values for IT7 and IT8.

例 2-3 已知公称尺寸为 20mm,求 IT7 和 IT8 的公差值。

Solution:
解:
The 20mm is within the size segment greater than 18~30mm:
20mm 在大于 18~30mm 的尺寸段内:

$$D = (18 \times 30)^{1/2} = 23.24mm$$

Numerical Values of Standard Tolerances
标准公差数值表

Tab. 2-2
表 2-2

Nominal size 公称尺寸		\multicolumn{20}{c}{Standard tolerance factor 标准公差等级}
>	≤	IT01	IT0	IT1	IT2	IT3	IT4	IT5	IT6	IT7	IT8	IT9	IT10	IT11	IT12	IT13	IT14	IT15	IT16	IT17	IT18
		\multicolumn{11}{c}{Standard tolerance 标准公差 (μm)}		\multicolumn{7}{c}{Standard tolerance 标准公差 (mm)}
—	3	0.3	0.5	0.8	1.2	2	3	4	6	10	14	25	40	60	0.1	0.14	0.25	0.4	0.6	1	1.4
3	6	0.4	0.6	1	1.5	2.5	4	5	8	12	18	30	48	75	0.12	0.18	0.3	0.48	0.5	1.2	1.8
6	10	0.4	0.6	1	1.5	2.5	4	6	9	15	22	36	58	90	0.15	0.22	0.36	0.58	0.9	1.5	2.2
10	18	0.5	0.8	1.2	2	3	5	8	11	18	27	43	70	110	0.18	0.27	0.43	0.7	1.1	1.8	2.7
18	30	0.6	1	1.5	2.5	4	6	9	13	21	33	52	84	130	0.21	0.33	0.52	0.84	1.3	2.1	3.3
30	50	0.6	1	1.5	2.5	4	7	11	16	25	39	62	100	160	0.25	0.39	0.62	1	1.6	2.5	3.9
50	80	0.8	1.2	2	3	5	8	13	19	30	46	74	120	190	0.3	0.46	0.74	1.2	1.9	3	4.6
80	120	1	1.5	2.5	4	6	10	15	22	35	54	87	140	220	0.35	0.54	0.87	1.4	2.2	3.5	5.4
120	180	1.2	2	3.5	5	8	12	18	25	40	63	100	160	250	0.4	0.63	1	1.6	2.5	4	6.3
180	250	2	3	4.5	7	10	14	20	29	46	72	115	185	290	0.46	0.72	1.15	1.85	2.9	4.6	7.2
250	315	2.5	4	6	8	12	16	23	32	52	81	130	210	320	0.52	0.81	1.3	2.1	3.2	5.2	8.1
315	400	3	5	7	9	13	18	25	36	57	89	140	230	360	0.57	0.89	1.4	2.3	3.6	5.7	8.9
400	500	4	6	8	10	15	20	27	40	63	97	155	250	400	0.63	0.97	1.55	2.5	4	6.3	9.7
500	630	—	—	9	11	16	22	32	44	70	110	175	280	440	0.7	1.1	1.75	2.8	4.4	7	11
630	800	—	—	10	13	18	25	36	50	80	125	200	320	500	0.8	1.25	2	3.2	5	8	12.5
800	1000	—	—	11	15	21	28	40	56	90	140	230	360	560	0.9	1.4	2.3	3.6	5.6	9	14
1000	1250	—	—	13	18	24	33	47	66	105	165	260	420	660	1.05	1.65	2.6	4.2	6.6	10.5	16.5
1250	1600	—	—	15	21	29	39	55	78	125	195	310	500	780	1.25	1.95	3.1	5	7.8	12.5	19.5
1600	2000	—	—	18	25	35	46	65	92	150	230	370	600	920	1.5	2.3	3.7	6	9.2	15	23
2000	2500	—	—	22	30	41	55	78	110	175	280	440	700	1110	1.75	2.8	4.4	7	11	17.5	28
2500	3150	—	—	26	36	50	68	96	135	210	330	540	860	1350	2.1	3.3	5.4	8.6	13.5	21	33

Note: the nominal sizes are in millimeters, and when the nominal sizes less than or equal to 1mm, there is no IT14 ~ IT18.

注:公称尺寸单位为 mm,公称尺寸小于或等于1mm 时,无 IT14~IT18。

Standard tolerance factor can be obtained from Equation (2-19):
由式(2-19)得标准公差因子：
$$i = 0.45D^{1/3} + 0.001D = 0.45 \times (23.24)^{1/3} + 0.001 \times 23.24 = 1.31 \mu m$$
It can be seen from Tab. 2-1 that IT7 = 16i and IT8 = 25i, i. e.
由表 2-1 查得 IT7 = 16i, IT8 = 25i, 即
$$IT7 = 16i = 16 \times 1.31 = 20.96 \mu m \approx 21 \mu m$$
$$IT8 = 25i = 25 \times 1.31 = 32.75 \mu m \approx 33 \mu m$$
Values can also be found directly from Tab. 2-2.
也可以直接从表 2-2 中查取数值。

2.2.3 Fundamental Deviation Series 基本偏差系列

The fundamental deviation is the upper limit deviation or the lower limit deviation of the part tolerance zone relative to the zero line, which is generally the deviation near the zero line. The basic deviation is the parameter to determine the position of tolerance zone, and it is the only index to standardize the position of tolerance zone.

基本偏差是确定零件公差带相对零线位置的上极限偏差或下极限偏差，一般为靠近零线的偏差。基本偏差是决定公差带位置的参数，它是公差带位置标准化的唯一指标。

2.2.3.1 Symbols and Characteristics of Fundamental Deviations 基本偏差代号及其特点

There are 28 fundamental deviations in total for holes and shafts, respectively. These different basic deviations constitute a series of basic deviations. All of them are represented by Latin letters. Upper case represents holes and lower case represents shafts. To avoid confusion, the letters I(i), L(l), O(o), Q(q) and W(w) are not used. And some fundamental deviations use double letters CD(cd), EF(ef), FG(fg), JS(js), ZA(za), ZB(zb) and ZC(zc) to meet some requirements. The fundamental deviation for a shaft is symmetry, in relation to the zero line, to the corresponding to the fundamental deviations of hole with same letters. The distribution of fundamental deviations for holes and shafts are shown in Fig. 2-8.

国标规定了孔和轴各有 28 种基本偏差，这些不同的基本偏差便构成了基本偏差系列。基本偏差的代号用拉丁字母表示，大写表示孔，小写表示轴。26 个字母中去掉 5 个易与其他参数相混淆的字母：I(i)、L(l)、O(o)、Q(q)、W(w)。为满足某些配合的需要，又增加了 7 个双写字母：CD(cd)、EF(ef)、FG(fg)、JS(js)、ZA(za)、ZB(zb)、ZC(zc)。相对于零线，轴的基本偏差与具有相同字母的孔的基本偏差对称。即得孔、轴各 28 个基本偏差代号，它们的分布如图 2-8 所示。

Characteristics of fundamental deviations of holes are as follows：
孔的基本偏差有以下主要特点：

A to G: the tolerance zones are totally above the zero line, the fundamental deviations are lower deviation EI, the value is positive (+).

A 到 G：公差带在零线上方，基本偏差为下极限偏差 EI，基本偏差值为正值。

H: the tolerance zone is totally above the zero line, the fundamental deviation is lower deviation,

EI = 0. At this time the hole is the datum hole.

H：公差带在零线上方，基本偏差为下极限偏差 EI，EI = 0，此时孔为基准孔。

JS：the tolerance zone is symmetry about the zero line, the fundamental deviation is the lower deviation or upper deviation. The absolute value of the limits deviation equals to half of the tolerance value.

JS：公差带完全对称于零线，基本偏差可以是上极限偏差亦可以是下极限偏差，基本偏差值的绝对值为公差值的一半。

J to ZC：the fundamental deviations are upper deviation ES, most value are negative.

J 到 ZC：基本偏差为上极限偏差 ES，基本偏差值多为负值。

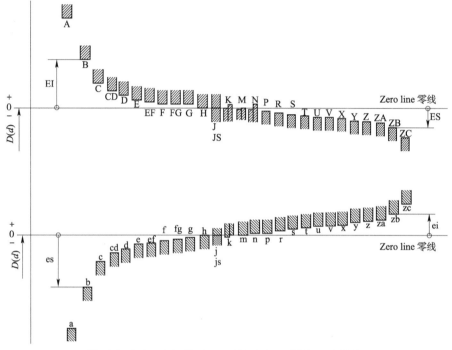

Fig. 2-8　Series of the Fundamental Deviations of Holes and Shafts
图 2-8　孔轴的基本偏差系列

Characteristics of fundamental deviations of shafts：

轴的基本偏差有以下主要特点：

a to g：the tolerance zones are totally below the zero line, the fundamental deviations are upper deviation es, the value is negative (−).

a 到 g：公差带在零线下方，基本偏差为上极限偏差 es，基本偏差值为负值。

h：the tolerance zone is totally below the zero line, the fundamental deviation is upper deviation, es = 0. At this time the shaft is the datum shaft.

h：公差带在零线下方，基本偏差为上极限偏差 es，es = 0，此时轴为基准轴。

js：the tolerance zone is symmetry about the zero line, the fundamental deviation is the lower deviation or upper deviation. The absolute value of the limits deviation equals to half of the tolerance value.

js：公差带完全对称于零线，基本偏差可以是上极限偏差亦可以是下极限偏差，基本偏差值的绝对值为公差值的一半。

j to zc: the tolerance zone is above the zero line, the fundamental deviations are lower deviation ei, and the basic deviation is mostly positive.

j 到 zc:公差带在零线上方,基本偏差为下极限偏差 ei,基本偏差值多为正值。

2.2.3.2 Fundamental Deviation Values of Shafts and Holes 孔和轴的基本偏差值

The values of fundamental deviations for shaft are calculated based on the fits of hole-basis system. According to different fit requirements, a serie of equations are achieved through statistical analysis based on production practices and a large number of experiments. It needn't calculate the values when designing; the engineers just directly look up the table. The basic deviation of the shaft is obtained by looking up the table, and the other deviation of the shaft is calculated according to the basic deviation and the standard tolerance of the shaft. The values of fundamental deviations for shafts are given in Tab. 2-3.

基孔制配合下,计算确定轴的基本偏差值。根据不同的配合要求,在大量的生产实践经验基础上,通过统计分析确定了基本偏差的计算方程。在进行设计时不需要计算基本偏差数值,通过查表的方式直接获取。轴的基本偏差通过查表获得,轴的另一个偏差根据轴的基本偏差和标准公差计算获得。轴的基本偏差值见表 2-3。

The fundamental deviation of the hole is obtained by the conversion of the fundamental deviation of the shaft. The principle of conversion is based on two principles of national standards: process equivalence and homonymic fit. Process equivalence: in the standard fit of hole-basis system and shaft-basis system, the process equivalence of hole and the shaft should be ensured, that is, the machining difficulty of the hole and the shaft is equal. Homonymic fit: using the same letter to represent the tolerance zone formed by the fundamental deviation of the hole and the shaft. The fit formed according to the hole-basis system and the shaft-basis system are called homonymic fit. The homonymic fit that satisfies the process equivalence has the same fit properties, that is, the type of fit is the same and the limit clearance amount or the limit interference amount is the same. According to the principles above, the fundamental deviation of the hole is converted according to the following two rules, as shown in Fig. 2-9.

孔的基本偏差是由轴的基本偏差换算得到的。换算的原则是基于国家标准的两条原则:工艺等价和同名配合。工艺等价:标准的基孔制和基轴制配合中,应保证孔和轴的工艺等价,即孔和轴的加工难易程度相当。同名配合:用同一字母表示孔和轴的基本偏差所组成的公差带,按照基孔制形成的配合和按照基轴制形成的配合称为同名配合。满足工艺等价的同名配合,其配合性质相同,即配合种类相同且极限间隙量或极限过盈量相等。根据上述原则,孔的基本偏差按以下两种规则换算,如图 2-9 所示。

(1) General Rule.

(1) 通用规则。

The absolute values of the fundamental deviations of the hole and the shaft represented by the same letter are equal, and the signs are opposite. The fundamental deviation of the hole is the reflection of the fundamental deviation of the shaft relative to the zero line, so it is also called the "reflection rule". The general rules apply in the following cases.

Numerical Values of the
轴的基本

Nominal size 公称尺寸 (mm)	a[①]	b[①]	c	cd	d	e	ef	f	fg	g	h	js[②]	j			k	
	Upper limit deviation 上极限偏差(es)																
	All standard tolerance grades 适用所有公差等级											5, 6	7	8	4~7	≤3, >7
≤3	−270	−140	−60	−34	−20	−14	−10	−6	−4	−2	0		−2	−4	−6	0	0
>3~6	−270	−140	−70	−46	−30	−20	−14	−10	−6	−4	0		−2	−5		+1	0
>6~10	−280	−150	−80	−56	−40	−25	−18	−13	−8	−5	0		−2	−5		+1	0
>10~14	−290	−150	−95		−50	−32		−16		−6	0		−3	−6		+1	0
>14~18	−290	−150	−95		−50	−32		−16		−6	0		−3	−6		+1	0
>18~24	−300	−160	−110		−65	−40		−20		−7	0		−4	−8		+2	0
>24~30	−300	−160	−110		−65	−40		−20		−7	0		−4	−8		+2	0
>30~40	−310	−170	−120		−80	−50		−25		−9	0		−5	−10		+2	0
>40~50	−320	−180	−130		−80	−50		−25		−9	0		−5	−10		+2	0
>50~65	−340	−190	−140		−100	−60		−30		−10	0		−7	−12		+2	0
>65~80	−360	−200	−150		−100	−60		−30		−10	0		−7	−12		+2	0
>80~100	−380	−220	−170		−120	−72		−36		−12	0		−9	−15		+3	0
>100~120	−410	−240	−180		−120	−72		−36		−12	0		−9	−15		+3	0
>120~140	−460	−260	−200		−145	−85		−43		−14	0		−11	−18		+3	0
>140~160	−520	−280	−210		−145	−85		−43		−14	0		−11	−18		+3	0
>160~180	−580	−310	−230		−145	−85		−43		−14	0		−11	−18		+3	0
>180~200	−660	−340	−240		−170	−100		−50		−15	0		−13	−21		+4	0
>200~225	−740	−380	−260		−170	−100		−50		−15	0		−13	−21		+4	0
>225~250	−820	−420	−280		−170	−100		−50		−15	0		−13	−21		+4	0
>250~280	−920	−480	−300		−190	−110		−56		−17	0	±IT$_n$/2	−16	−26		+4	0
>280~315	−1050	−540	−330		−190	−110		−56		−17	0		−16	−26		+4	0
>315~355	−1200	−600	−360		−210	−125		−62		−18	0		−18	−28		+4	0
>355~400	−1350	−680	−400		−210	−125		−62		−18	0		−18	−28		+4	0
>400~450	−1500	−760	−440		−230	−135		−68		−20	0		−20	−32		+5	0
>450~500	−1650	−840	−480		−230	−135		−68		−20	0		−20	−32		+5	0
>500~560					−260	−145		−76		−22	0						0
>560~630					−260	−145		−76		−22	0						0
>630~710					−290	−160		−80		−24	0						0
>710~800					−290	−160		−80		−24	0						0
>800~900					−320	−170		−86		−26	0						0
>900~1000					−320	−170		−86		−26	0						0
>1000~1120					−350	−195		−98		−28	0						0
>1120~1250					−350	−195		−98		−28	0						0
>1250~1400					−390	−220		−110		−30	0						0
>1400~1600					−390	−220		−110		−30	0						0
>1600~1800					−430	−240		−120		−32	0						0
>1800~2000					−430	−240		−120		−32	0						0
>2000~2240					−480	−260		−130		−34	0						0
>2240~2500					−480	−260		−130		−34	0						0
>2500~2800					−520	−290		−145		−38	0						0
>2800~3150					−520	−290		−145		−38	0						0

Note: ① Fundamental deviations a and b shall not be used for basic sizes less than or equal to 1mm.

② For tolerance grades js7 to js11, if the IT value number, n, is an odd number, then the value = $\pm(IT-1)/n$.

注:①公称尺寸小于或等于1mm时,不使用基本偏差a和b。

②对于公差等级js7至js11,如果IT值n为奇数,则该值 = $\pm(IT-1)/n$。

Chapter 2 Size Tolerances and Fits 尺寸公差与配合

Fundamental Deviations of Shafts
偏差数值表

Tab. 2-3
表 2-3

m	n	p	r	s	t	u	v	x	y	z	za	zb	zc
\multicolumn{14}{c}{Lower limit deviation 下极限偏差（ei）}													
\multicolumn{14}{c}{All standard tolerance grades 适用所有公差等级}													
+2	+4	+6	+10	+14		+18		+20		+26	+32	+40	+80
+4	+8	+12	+15	+19		+23		+28		+35	+42	+50	+90
+6	+10	+15	+19	+23		+28		+34		+42	+52	+67	+97
+7	+12	+18	+23	+28		+33		+40		+50	+64	+90	+130
						+39	+45		+60	+77	+108	+150
+8	+15	+22	+28	+35		+41	+47	+54	+63	+73	+98	+136	+188
					+41	+48	+55	+64	+75	+88	+118	+160	+218
+9	+17	+26	+34	+43	+48	+60	+68	+80	+94	+112	+148	+200	+274
					+54	+70	+81	+97	+114	+136	+180	+242	+325
+11	+20	+32	+41	+53	+66	+87	+102	+122	+144	+172	+226	+300	+405
			+43	+59	+75	+102	+120	+146	+174	+210	+274	+360	+480
+13	+23	+37	+51	+71	+91	+124	+146	+178	+214	+258	+335	+445	+585
			+54	+79	+104	+144	+172	+210	+254	+310	+400	+525	+690
+15	+27	+43	+63	+92	+122	+170	+202	+248	+300	+365	+470	+620	+800
			+65	+100	+134	+190	+228	+280	+340	+415	+535	+700	+900
			+68	+108	+146	+210	+252	+310	+380	+465	+600	+780	+1000
+17	+31	+50	+77	+122	+166	+236	+284	+350	+425	+520	+670	+880	+1150
			+80	+130	+180	+258	+310	+385	+470	+575	+740	+960	+1250
			+84	+140	+196	+284	+340	+425	+520	+640	+820	+1050	+1350
+20	+34	+56	+94	+158	+218	+315	+385	+475	+580	+710	+920	+1200	+1550
			+98	+170	+240	+350	+425	+525	+650	+790	+1000	+1300	+1700
+21	+37	+62	+108	+190	+268	+390	+475	+590	+730	+900	+1150	+1500	+1900
			+114	+208	+294	+435	+530	+660	+820	+1000	+1300	+1650	+2100
+23	+40	+68	+126	+232	+330	+490	+595	+740	+920	+1100	+1450	+1850	+2400
			+132	+252	+360	+540	+660	+820	+1000	+1250	+1600	+2100	+2600
+26	+44	+78	+150	+280	+400	+600							
			+155	+310	+450	+660							
+30	+50	+88	+175	+340	+500	+740							
			+185	+380	+560	+840							
+34	+56	+100	+210	+430	+620	+940							
			+220	+473	+680	+1050							
+40	+66	+120	+250	+520	+780	+1150							
			+260	+580	+840	+1300							
+48	+78	+140	+300	+640	+960	+1450							
			+330	+720	+1050	+1600							
+58	+92	+170	+370	+820	+1200	+1850							
			+400	+920	+1350	+2000							
+68	+110	+195	+440	+1000	+1500	+2300							
			+460	+1100	+1650	+2500							
+76	+135	+240	+550	+1250	+1900	+2900							
			+580	+1400	+2100	+3200							

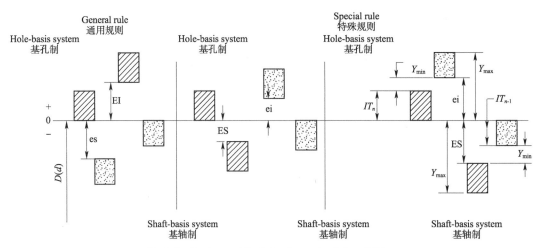

Fig. 2-9 Fundamental Deviation Conversion Rules of Holes

图 2-9 孔的基本偏差换算规则

用同一字母表示的孔、轴的基本偏差的绝对值相等,符号相反。孔的基本偏差是轴的基本偏差相对于零线的倒影,因此又称为"倒影规则"。通用规则适用于以下情况。

For A to H, since the absolute values of the fundamental deviation EI of the hole and the fundamental deviation es of the corresponding shaft are equal to the minimum clearance, the conversion principle is EI = − es.

对于 A ~ H,因其基本偏差 EI 与对应轴的基本偏差 es 的绝对值都等于最小间隙,故其换算原则为:EI = − es。

For K to ZC, the cases is applicable for fundamental deviations K, M and N in standard tolerance grades up to IT8, and deviations from P to ZC in standard tolerance grades up to IT7. The conversion principle is ES = − ei. With the exception of the nominal size over 3mm for fundamental deviation N in standard tolerance grades up to IT8, the fundamental deviation is ES = 0.

对于 K ~ ZC,标准公差大于 IT8 的 K、M、N 和大于 IT7 的 P ~ ZC,换算原则为:ES = − ei。标准公差大于 IT8、公称尺寸大于 3mm 的 N 除外,其基本偏差 ES = 0。

(2)Special Rule.

(2)特殊规则。

When using the same letter to represent the fundamental deviation of the hole and the shaft, the fundamental deviations ES of the hole and the fundamental deviation ei of the shaft are opposite, and the absolute value differs by a Δ value.

用同一字母表示孔、轴基本偏差时,孔的基本偏差 ES 和轴的基本偏差 ei 符号相反,而绝对值相差一个 Δ 值。

Because it is more difficult to process the hole than the shaft of the same tolerance grade in the higher tolerance grades, the hole fit of one grade lower than the shaft is often used, that is, the different fit, and the fit properties formed by the two fit systems are required to be the same.

因为在较高的公差等级中,同一公差等级的孔比轴加工困难,因而常采用比轴低一级的孔相配合,即异级配合,并要求两种配合制所形成的配合性质相同。

For the hole-basis fit:
基孔制配合时:

$$Y_{\min} = ES - ei = +IT_n - ei \tag{2-21}$$

For the shaft-basis fit
基轴制配合时:

$$Y_{\min} = ES - ei = ES - (-IT_{n-1}) \tag{2-22}$$

Following the rule, the minimum interference should be equal:
要求具有相同的配合性质,故有:

$$+IT_n - ei = ES - (-IT_{n-1}) \tag{2-23}$$

The basic deviation of the hole is obtained:
由此得出孔的基本偏差为:

$$ES = -ei + \Delta, \Delta = IT_n - IT_{n-1} \tag{2-24}$$

IT_n is the standard tolerance of a hole of a certain grade, and IT_{n-1} is the standard tolerance of a shaft one grade higher than a hole of a certain grade.

IT_n 为某一级孔的标准公差,IT_{n-1} 为比某一级孔高一级的轴的标准公差。

The special cases are only applicable for the nominal size from 3mm up to 500mm (including) for fundamental deviations K, M, and N in standard tolerance grades up to and including IT8, and deviations P to ZC in standard tolerance grades up to and including IT7.

特殊规则适用于以下情况:公称尺寸大于3mm 小于或等于500mm,标准公差小于或等于IT8 的 J、K、M、N 和标准公差小于或等于IT7 的 P~ZC。

The other deviation of the hole should be calculated and determined according to the basic deviation and the standard tolerance of the hole.

孔的另一个偏差,根据孔的基本偏差和标准公差计算确定。

Besides the method of calculating, the basic deviation of the hole can also be obtained by checking the table directly. The basic deviation values of the hole are shown in Tab. 2-4.

孔的基本偏差除了计算外,还可以直接查表获得。孔的基本偏差值见表2-4。

Example 2-4 Check the table to determine the limit deviation of the hole and the shaft in the fitting of $\phi30H8/f7$ and $\phi30F8/h7$, calculate the limit clearance of two pairs of fits and draw the tolerance zone diagram.

例 2-4 查表确定 $\phi30H8/f7$ 和 $\phi30F8/h7$ 配合中孔、轴的极限偏差,计算两对配合的极限间隙并绘制公差带图。

Solution:
解:

(1) Check the table to determine the limit deviation of the hole and the shaft in the fitting of $\phi30h8/F7$.

(1) 查表确定 $\phi30H8/f7$ 配合中的孔与轴的极限偏差。

The nominal size of $\phi30$ is in the size segment >18~30mm. According to Tab. 2-2:
公称尺寸 $\phi30$ 属于>18~30 mm 尺寸段,由表2-2 得:

$$IT7 = 21\mu m, IT8 = 33\mu m$$

Numerical Values of the
孔的基本

Nominal size 公称尺寸 (mm)	A[①]	B[①]	C	CD	D	E	EF	F	FG	G	H	JS[②]	J 6	J 7	J 8	K ≤8	K >8	M ≤8	M >8
				Lower limit deviation 下极限偏差(EI)															
				All standard tolerance grades 适用所有公差等级															
≤3	+270	+140	+60	+34	+20	+14	+10	+6	+4	+2	0		+2	+4	+6	0	0	−2	−2
>3~6	+270	+140	+70	+46	+30	+20	+14	+10	+6	+4	0		+5	+6	+10	−1+Δ		−4+Δ	−4
>6~10	+280	+150	+80	+56	+40	+25	+18	+13	+8	+5	0		+5	+8	+12	−1+Δ		−6+Δ	−6
>10~14	+290	+150	+95		+50	+32		+16		+6	0		+6	+10	+15	−1+Δ		−7+Δ	−7
>14~18	+290	+150	+95		+50	+32		+16		+6	0		+6	+10	+15	−1+Δ		−7+Δ	−7
>18~24	+300	+160	+110		+65	+40		+20		+7	0		+8	+12	+20	−2+Δ		−8+Δ	−8
>24~30	+300	+160	+110		+65	+40		+20		+7	0		+8	+12	+20	−2+Δ		−8+Δ	−8
>30~40	+310	+170	+120		+80	+50		+25		+9	0		+10	+14	+24	−2+Δ		−9+Δ	−9
>40~50	+320	+180	+130		+80	+50		+25		+9	0		+10	+14	+24	−2+Δ		−9+Δ	−9
>50~65	+340	+190	+140		+100	+60		+30		+10	0		+13	+18	+28	−2+Δ		−11+Δ	−11
>65~80	+360	+200	+150		+100	+60		+30		+10	0		+13	+18	+28	−2+Δ		−11+Δ	−11
>80~100	+380	+220	+170		+120	+72		+36		+12	0		+16	+22	+34	−3+Δ		−13+Δ	−13
>100~120	+410	+240	+180		+120	+72		+36		+12	0		+16	+22	+34	−3+Δ		−13+Δ	−13
>120~140	+460	+260	+200		+145	+85		+43		+14	0	±IT$_n$/2	+18	+26	+41	−3+Δ		−15+Δ	−15
>140~160	+520	+280	+210		+145	+85		+43		+14	0		+18	+26	+41	−3+Δ		−15+Δ	−15
>160~180	+580	+310	+230		+145	+85		+43		+14	0		+18	+26	+41	−3+Δ		−15+Δ	−15
>180~200	+660	+340	+240		+170	+100		+50		+15	0		+22	+30	+47	−4+Δ		−17+Δ	−17
>200~225	+740	+380	+260		+170	+100		+50		+15	0		+22	+30	+47	−4+Δ		−17+Δ	−17
>225~250	+820	+420	+280		+170	+100		+50		+15	0		+22	+30	+47	−4+Δ		−17+Δ	−17
>250~280	+920	+480	+300		+190	+110		+56		+17	0		+25	+36	+55	−4+Δ		−20+Δ	−20
>280~315	+1050	+540	+330		+190	+110		+56		+17	0		+25	+36	+55	−4+Δ		−20+Δ	−20
>315~355	+1200	+600	+360		+210	+125		+62		+18	0		+29	+39	+60	−4+Δ		−21+Δ	−21
>355~400	+1350	+680	+400		+210	+125		+62		+18	0		+29	+39	+60	−4+Δ		−21+Δ	−21
>400~450	+1500	+760	+440		+230	+135		+68		+20	0		+33	+43	+66	−5+Δ		−23+Δ	−23
>450~500	+1650	+840	+480		+230	+135		+68		+20	0		+33	+43	+66	−5+Δ		−23+Δ	−23
>500~560					+260	+145		+76		+22	0					0		−26	
>560~630					+260	+145		+76		+22	0					0		−26	
>630~710					+290	+160		+80		+24	0					0		−30	
>710~800					+290	+160		+80		+24	0					0		−30	
>800~900					+320	+170		+86		+26	0					0		−34	
>900~1000					+320	+170		+86		+26	0					0		−34	
>1000~1120					+350	+195		+98		+28	0					0		−40	
>1120~1250					+350	+195		+98		+28	0					0		−40	
>1250~1400					+390	+220		+110		+30	0					0		−48	
>1400~1600					+390	+220		+110		+30	0					0		−48	
>1600~1800					+430	+240		+120		+32	0					0		−58	
>1800~2000					+430	+240		+120		+32	0					0		−58	
>2000~2240					+480	+260		+130		+34	0					0		−68	
>2240~2500					+480	+260		+130		+34	0					0		−68	
>2500~2800					+520	+290		+145		+38	0					0		−76	
>2800~3150					+520	+290		+145		+38	0					0		−76	

Note:①Fundamental deviations A and B for all grades and N for grades above IT8 shall not be used for basic sizes less than or equal to 1mm.

②For tolerance grades JS7 to JS11, if the IT value number, n, is an odd number, then the value = ±(IT−1)/n.

注:①所有等级的基本偏差 A 和 B 以及 IT8 以上等级的基本偏差 N 不得用于小于或等于1mm 的公称尺寸。

②对于公差等级 JS7 至 JS11,如果 IT 值 n 为奇数,则该值 = ±(IT−1)/n。

Chapter 2　Size Tolerances and Fits 尺寸公差与配合

Fundamental Deviations of Holes
偏差数值表

Tab. 2-4
表2-4

N		P~ZC	P	R	S	T	U	V	X	Y	Z	ZA	ZB	ZC	Values of Δ					
			Upper limit deviation 上极限偏差(ES)												IT3	IT4	IT5	IT6	IT7	IT8
≤8	>8	≤7					>7													
−4			−6	−10	−14		−18		−20		−26	−32	−40	−80	0					
−8+Δ	0		−12	−15	−19		−23		−28		−35	−42	−50	−90	1	1.5	1	3	4	6
−10+Δ	0		−15	−19	−23		−28		−34		−42	−52	−67	−97	1	1.5	2	3	6	7
−12+Δ	0		−18	−23	−28		−33		−40		−50	−64	−90	−130	1	2	3	3	7	9
									−45		−60	−77	−108	−150						
−15+Δ	0		−22	−28	−35		−41	−47	−54	−63	−73	−98	−136	−188	1.5	2	3	4	8	12
						−41	−48	−55	−64	−75	−88	−118	−160	−218						
−17+Δ	0		−26	−34	−43	−48	−60	−68	−80	−94	−112	−148	−200	−274	1.5	3	4	5	9	14
						−54	−70	−81	−97	−114	−136	−180	−242	−325						
−20+Δ	0		−32	−41	−53	−66	−87	−102	−122	−144	−172	−226	−300	−405	2	3	5	6	11	16
				−43	−59	−75	−102	−120	−146	−174	−210	−274	−360	−480						
−23+Δ	0		−37	−51	−71	−91	−124	−146	−178	−214	−258	−335	−445	−585	2	4	5	7	13	19
				−54	−79	−104	−144	−172	−210	−254	−310	−400	−525	−690						
−27+Δ	0	Values as for standard grades above IT7 increase Δ IT7以上等级的数值增加Δ	−43	−63	−92	−122	−170	−202	−248	−300	−365	−470	−620	−800	3	4	6	7	15	23
				−65	−100	−134	−190	−228	−280	−340	−415	−535	−700	−900						
				−68	−108	−146	−210	−252	−310	−380	−465	−600	−780	−1000						
−31+Δ	0		−50	−77	−122	−166	−236	−284	−350	−425	−520	−670	−880	−1150	3	4	6	9	17	26
				−80	−130	−180	−258	−310	−385	−470	−575	−740	−960	−1250						
				−84	−140	−196	−284	−340	−425	−520	−640	−820	−1050	−1350						
−34+Δ	0		−56	−94	−158	−218	−315	−385	−475	−580	−710	−920	−1200	−1550	4	4	7	9	20	29
				−98	−170	−240	−350	−425	−525	−650	−790	−1000	−1300	−1700						
−37+Δ	0		−62	−108	−190	−268	−390	−475	−590	−730	−900	−1150	−1500	−1900	4	5	7	11	21	32
				−114	−208	−294	−435	−530	−660	−820	−1000	−1300	−1650	−2100						
−40+Δ	0		−68	−126	−232	−330	−490	−595	−740	−920	−1100	−1450	−1850	−2400	5	5	7	13	23	34
				−132	−252	−360	−540	−660	−820	−1000	−1250	−1600	−2100	−2600						
−44			−78	−150	−280	−400	−600													
				−155	−310	−450	−660													
−50			−88	−175	−340	−500	−740													
				−185	−380	−560	−840													
−56			−100	−210	−430	−620	−940													
				−220	−473	−680	−1050													
−66			−120	−250	−520	−780	−1150													
				−260	−580	−840	−1300													
−78			−140	−300	−640	−960	−1450													
				−330	−720	−1050	−1600													
−92			−170	−370	−820	−1200	−1850													
				−400	−920	−1350	−2000													
−110			−195	−440	−1000	−1500	−2300													
				−460	−1100	−1650	−2500													
−135			−240	−550	−1250	−1900	−2900													
				−580	−1400	−2100	−3200													

For the datum hole H8, EI = 0, ES is：

对于基准孔 H8 的 EI = 0,其 ES 为：

$$ES = EI + IT8 = +33\mu m$$

For f7, according to Tab. 2-3, es = −20μm, ei is：

对于 f7,查表 2-3 得 es = −20μm,其 ei 为：

$$ei = es - IT7 = -20 - 21 = -41\mu m$$

Thus it can be obtained that：

由此可得：

$\phi 30H8 = \phi 30^{+0.033}_{0}$, $\phi 30f7 = \phi 30^{-0.020}_{-0.041}$。

(2) Check the table to determine the limit deviation of $\phi 30F8/h7$ with the hole and the shaft.

(2) 查表确定 $\phi 30F8/h7$ 配合中孔与轴的极限偏差。

For F8, according to Tab. 2-4, EI = +20μm, ES is：

对于 F8,由表 2-4 得 EI = +20μm,其 ES 为：

$$ES = EI + IT8 = +20 + 33 = +53\mu m$$

For the datum shaft h7, es = 0, ei is：

对基准轴 h7 的 es = 0,其 ei 为：

$$ei = es - IT7 = -21\mu m$$

Thus it can be obtained that：

由此可得：

$\phi 30F8 = \phi 30^{+0.053}_{+0.020}$, $\phi 30h7 = \phi 30^{0}_{-0.021}$。

(3) The limit clearance between $\phi 30H8/f7$ and $\phi 30F8/h7$ is calculated.

(3) 计算 $\phi 30H8/f7$ 和 $\phi 30F8/h7$ 配合的极限间隙。

For $\phi 30H8/f7$：

对于 $\phi 30H8/f7$：

$$X_{max} = ES - ei = +33 - (-41) = +74\mu m$$
$$X_{min} = EI - es = 0 - (-20) = +20\mu m$$

For $\phi 30F8/h7$：

对于 $\phi 30F8/h7$：

$$X'_{max} = ES - ei = +53 - (-21) = +74\mu m$$
$$X'_{min} = EI - es = +20 - 0 = +20\mu m$$

(4) The tolerance zone diagrams are drawn using the limit deviation and limit clearance values calculated above, as shown in Fig. 2-10.

(4) 用上面计算的极限偏差和极限间隙值绘制公差带图,如图 2-10 所示。

As can be seen from the above calculation and Fig. 2-10, the $\phi 30H8/f7$ and $\phi 30F8/h7$ fits have the same maximum and minimum clearance, that is, the fit properties are the same.

由上述计算和图 2-10 可见,$\phi 30H8/f7$ 和 $\phi 30F8/h7$ 两对配合的最大间隙和最小间隙均相等,即配合性质相同。

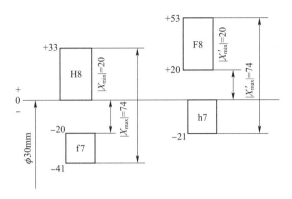

Fig. 2-10 φ30H8/f7 and φ30F8/h7 Tolerance Zone
图 2-10 φ30H8/f7 和 φ30F8/h7 的公差带

Example 2-5 φ20 in the dimensoin segment >18~30, known IT6 = 13μm, IT7 = 21μm, the basic deviation of φ20k6 is the lower limit deviation, and ei = +2μm. Try not using the tables, determine the φ20H7/k6 and φ20K7/h6 combinations of the hole and the shaft limit deviation, calculate the limit clearance or interference, and draw the tolerance zone diagrams.

例 2-5 φ20 在尺寸段 >18~30,已知 IT6 = 13μm,IT7 = 21μm,φ20k6 的基本偏差是下极限偏差,且 ei = +2μm。试不用查表法,确定 φ20H7/k6 和 φ20K7/h6 两种配合的孔、轴极限偏差,计算极限间隙或过盈,并绘制公差带图。

Solution:
解:

If you calculate the limit deviation, you must know the standard tolerance and the basic deviation. Here, the standard tolerances are known, so the four basic deviations of the two fits should be solved.

要求极限偏差,就必须知道标准公差和基本偏差。这里,标准公差是已知的,所以求出 2 个配合的 4 个基本偏差就行了。

(1) Determine φ20H7/k6.
(1) 确定 φ20H7/k6。
Start with reference hole φ20H7, for grade 7 reference hole:
从基准孔 φ20H7 开始,对于 7 级基准孔:

$$EI = 0, ES = EI + IT7 = +21\mu m$$

So the limit deviation of the reference hole is:
所以基准孔的极限偏差为:

$$\phi 20^{+0.021}_{0}$$

For φ20k6, ei = +2μm is known, and another limit deviation is obtained. The tolerance zone is above the zero line, then:
对于 φ20k6,已知 ei = +2μm,求另一极限偏差。公差带在零线以上,则有:

$$es = ei + IT6 = +2 + 13 = +15\mu m$$

So the limit deviation of the shaft is:
所以轴的极限偏差为:

$$\phi 20^{+0.015}_{+0.002}$$

The code name of the fit is:
配合代号为:

$$\phi 20 \frac{\mathrm{H7}\binom{+0.021}{0}}{\mathrm{k6}\binom{+0.015}{+0.002}}$$

Tolerance zones overlap, that is, it's a transition fit.
公差带交叠,过渡配合。

Limit clearance or interference:
极限间隙或过盈:

$$X_{\max} = \mathrm{ES} - \mathrm{ei} = +0.021 - (+0.002) = +0.019\mu\mathrm{m}$$
$$Y_{\max} = \mathrm{EI} - \mathrm{es} = 0 - (+0.015) = -0.015\mu\mathrm{m}$$

(2) Determine $\phi 20\mathrm{K7/h6}$.
(2) 确定 $\phi 20\mathrm{K7/h6}$。

Start with reference shaft $\phi 20\mathrm{h6}$, for the grade 6 reference shaft:
从基准轴 $\phi 20\mathrm{h6}$ 开始,对于 6 级基准轴:

$$\mathrm{es} = 0, \mathrm{ei} = \mathrm{es} - \mathrm{IT6} = -0.013\mu\mathrm{m}$$

So the limit deviation of the shaft is:
所以基准轴的极限偏差为:

$$\phi 20^{\ 0}_{-0.013}$$

$\phi 20\mathrm{K7}$ is a grade 7 hole, transition fit, the standard tolerance is less than grade 8, so it is opposite to the basic deviation sign of the corresponding shaft k7 of the hole, the absolute value differs a Δ value. Since the basic deviation is independent of the tolerance grade, the basic deviation of k7 is the same as that of k6.

$\phi 20\mathrm{K7}$ 是 7 级孔,过渡配合,标准公差小于 8 级,故与该孔对应轴 k7 的基本偏差符号相反,绝对值相差一个 Δ 值。由于基本偏差与公差等级无关,k7 的基本偏差与 k6 的基本偏差一样。

The basic deviation of K7 is the upper limit deviation, then:
K7 的基本偏差为上极限偏差,则有:

$$\Delta = \mathrm{IT7} - \mathrm{IT6} = 21 - 13 = 8\mu\mathrm{m}$$

The fundamental deviation of k6 is the lower limit deviation, $\mathrm{ei} = +2\mu\mathrm{m}$, so the fundamental deviation of k7 is $\mathrm{ei} = +2\mu\mathrm{m}$, then:
k6 基本偏差是下极限偏差,且 $\mathrm{ei} = +2\mu\mathrm{m}$,所以,k7 的基本偏差为 $\mathrm{ei} = +2\mu\mathrm{m}$,则有:

$$\mathrm{ES} = -\mathrm{ei} + \Delta = -(+2) + 8 = +6\mu\mathrm{m}$$

Another limit deviation of the hole is:
孔的另一个极限偏差是:

$$\mathrm{EI} = \mathrm{ES} - \mathrm{IT7} = +6 - 21 = -15\mu\mathrm{m}$$

Therefore, the limit deviation of the hole is:
所以该孔的极限偏差为:

$$\phi 20^{+0.006}_{-0.015}$$

The code name of the fit is:
配合代号是:

$$\phi 20 \frac{K7\left(^{+0.006}_{-0.015}\right)}{h6\left(^{0}_{-0.013}\right)}$$

Tolerance zones overlap, that is, it's a transition fit.

公差带交叠,过渡配合。

Limit clearance or interference:

极限间隙或过盈:

$$X'_{max} = ES - ei = +0.006 - (-0.013) = +0.019 \mu m$$
$$Y'_{max} = EI - es = (-0.015) - 0 = -0.015 \mu m$$

According to the calculations above, $\phi 20H7/k6$ and $\phi 20K7/h6$ fit have the same maximum clearance and maximum interference, that is, the fit properties are equal.

由上述计算可知,$\phi 20H7/k6$ 和 $\phi 20K7/h6$ 两个配合的最大间隙和最大过盈相等,即配合性质相等。

(3) Draw a tolerance zone diagram, as shown in Fig. 2-11.

(3) 画公差带图,如图 2-11 所示。

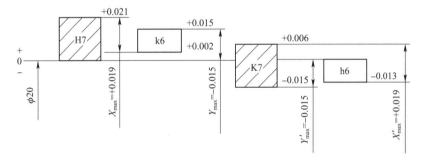

Fig. 2-11 $\phi 20H7/k6$ and $\phi 20K7/h6$ Tolerance Zone

图 2-11 $\phi 20H7/k6$ 和 $\phi 20K7/h6$ 的公差带

2.2.4 Standardization of Tolerance Grade and Fits 公差带与配合的标准化

In the national standards of tolerances and fits, there are 28 fundamental deviations and 20 tolerance grades for nominal size up to 500mm (including) and 18 tolerance grades for nominal size from 500mm up to 3150mm. A large number of tolerance grades of different sizes and positions can be obtained by combining any basic deviation with any standard tolerance. For the nominal size up to 500mm (including), there are 543 tolerance zones for holes and 544 tolerance zones for shafts. These tolerance zones can compose 295392 fits. So many tolerance zones and fits can meet all trades and professions, however, it is clearly not economical to use all the tolerance zones and fits. So the general, common and preferred tolerance zones and fits are selected.

国家标准中提供了 28 种基本偏差系列、20 种标准公差系列(公称尺寸小于或等于 500mm)和 18 种标准公差系列(公称尺寸介于 500mm 和 3150mm 之间),将任意基本偏差与任意标准公差组合,可以得到大小与位置不同的大量公差带。在公称尺寸≤500mm 范围内,孔的公差带有 543 个，轴的公差带有 544 个，可以组成 295392 个配合形式。显然，这么多的公差带与配合都使用是不经济的，因此需要挑选出一般、常用和优先公差带与配合。

Fig. 2-12 and Fig. 2-13 show the general, common and preferred tolerance zones for shafts and holes respectively which the nominal sizes up to 500mm (including). The tolerance zones in boxes are common tolerance zones, and that in the circles are preferred tolerance zones.

孔和轴的一般、常用和优先的公差带分别如图 2-12 和图 2-13 所示。方框中的公差带为常用公差带，圆框中的公差带为优先公差带。

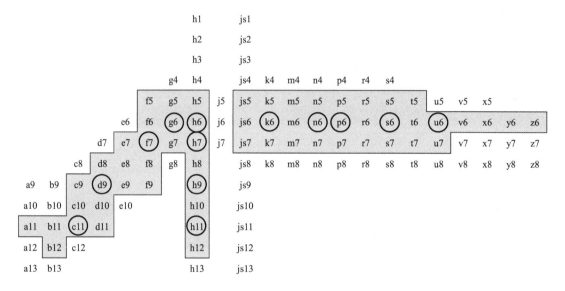

Fig. 2-12　General, Common and Preferred Tolerance Zones for Shafts

图 2-12　轴的一般、常用和优先公差带

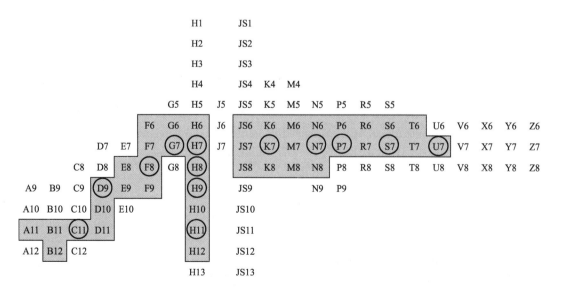

Fig. 2-13　General, Common and Preferred Tolerance Zones for Holes

图 2-13　孔的一般、常用和优先公差带

Chapter 2　Size Tolerances and Fits 尺寸公差与配合

On the basis of the above recommended tolerance zones for shafts and holes, the national standard also recommends the combination of tolerance zones for shafts and holes. Tab. 2-5 and Tab. 2-6 show the common and preferred fits for hole-basis fits and shaft-basis fits respectively which the nominal size up to 500mm (including). The fits marked with circles are preferred fits.

在上述推荐的轴、孔公差带的基础上，国家标准还推荐了孔、轴公差带的组合。公称尺寸小于或等于 500mm 时，表 2-5 和表 2-6 对基孔制和基轴制分别规定了常用与优先配合，圆圈中为优先配合。

Common and Preferred Fits for Hole-basis System　　Tab. 2-5
基孔制常用和优先配合　　表 2-5

Basic hole 基准孔	\multicolumn{20}{c}{Shaft 轴}
	a	b	c	d	e	f	g	h	js	k	m	n	p	r	s	t	u	v	x	y	z
	\multicolumn{8}{c}{Clearance fits 间隙配合}	\multicolumn{3}{c}{Transition fits 过渡配合}	\multicolumn{9}{c}{Interference fits 过盈配合}
H6						H6/f5	H6/g5	H6/h5	H6/js5	H6/k5	H6/m5	H6/n5	H6/p5	H6/r5	H6/s5	H6/t5					
H7						H7/f6	(H7/g6)	(H7/h6)	H7/js6	(H7/k6)	H7/m6	(H7/n6)	(H7/p6)	H7/r6	(H7/s6)	H7/t6	(H7/u6)	H7/v6	H7/x6	H7/y6	H7/z6
H8					H8/e7	(H8/f7)	H8/g7	(H8/h7)	H8/js7	H8/k7	H8/m7	H8/n7	H8/p7	H8/r7	H8/s7	H8/t7	H8/u7				
				H8/d8	H8/e8	H8/f8		H8/h8													
H9			H9/c9	(H9/d9)	H9/e9	H9/f9		(H9/h9)													
H10			H10/c10	H10/d10				H10/h10													
H11	H11/a11	H11/b11	(H11/c11)	H11/d11				(H11/h11)													
H12		H12/b12						H12/h12													

Common and Preferred Fits for Shaft-basis System　　Tab. 2-6
基轴制常用和优先配合　　表 2-6

Basic shaft 基准轴	\multicolumn{20}{c}{Hole 孔}
	A	B	C	D	E	F	G	H	JS	K	M	N	P	R	S	T	U	V	X	Y	Z
	\multicolumn{8}{c}{Clearance fits 间隙配合}	\multicolumn{3}{c}{Transition fits 过渡配合}	\multicolumn{9}{c}{Interference fits 过盈配合}
h5						F6/h5	G6/h5	H6/h5	JS6/h5	K6/h5	M6/h5	N6/h5	P6/h5	R6/h5	S6/h5	T6/h5					
h6						F7/h6	(G7/h6)	(H7/h6)	JS7/h6	(K7/h6)	M7/h6	(N7/h6)	(P7/h6)	R7/h6	(S7/h6)	T7/h6	(U7/h6)				

Continued 续上表

Basic shaft 基准轴	Hole 孔																				
	A	B	C	D	E	F	G	H	JS	K	M	N	P	R	S	T	U	V	X	Y	Z
	Clearance fits 间隙配合								Transition fits 过渡配合				Interference fits 过盈配合								
h7					$\frac{E8}{h7}$	$\frac{F8}{h7}$	$\frac{G8}{h7}$	$\frac{H8}{h7}$	$\frac{JS8}{h7}$	$\frac{K8}{h7}$	$\frac{M8}{h7}$	$\frac{N8}{h7}$									
h8				$\frac{D8}{h8}$	$\frac{E8}{h8}$	$\frac{F8}{h8}$		$\frac{H8}{h8}$													
h9				$\frac{D9}{h9}$	$\frac{E9}{h9}$	$\frac{F9}{h9}$		$\frac{H9}{h9}$													
h10				$\frac{D10}{h10}$				$\frac{H10}{h10}$													
h11	$\frac{A11}{h11}$	$\frac{B11}{h11}$	$\frac{C11}{h11}$	$\frac{D11}{h11}$				$\frac{H11}{h11}$													
h12		$\frac{B12}{h12}$						$\frac{H12}{h12}$													

2.2.5 Untoleranced Tolerance 未注尺寸公差

General tolerance refers to the tolerance that can be guaranteed for the general machining capacity of machine tools and equipment under common process conditions in the workshop. *General tolerances—Tolerances for linear and angular sizes without individual tolerance indications* (GB/T 1804—2000) provides the general tolerance grades and their deviation values. Permissible deviations for linear sizes except for broken edges are shown in Tab. 2-7. Permissible deviations for broken edges (external radii and chamfer heights) are shown in Tab. 2-8.

未注公差是指在车间普通工艺条件下机床设备一般加工能力可保证的公差。《一般公差 未注公差的线性和角度尺寸的公差》(GB/T 1804—2000)对线性尺寸的未注公差进行了规定。未注公差的线性尺寸极限偏差数值见表2-7。倒圆半径与倒角高度尺寸极限偏差数值见表2-8。

Permissible Deviations for Linear Sizes Except for Broken Edges Tab. 2-7
未注公差的线性尺寸极限偏差数值 表2-7

Tolerance grade 公差等级		Permissible deviations for nominal size range 尺寸分段的允许偏差(mm)							
Designation 等级	Description 属性	0.5~3	>3~6	>6~30	>30~120	>120~400	>400~1000	>1000~2000	>2000~4000
f	Fine 精密	±0.05	±0.05	±0.1	±0.015	±0.2	±0.3	±0.5	—
m	Medium 中等	±0.1	±0.1	±0.2	±0.3	±0.5	±0.8	±1.2	±2

Chapter 2 Size Tolerances and Fits 尺寸公差与配合

Continued 续上表

Tolerance grade 公差等级		Permissible deviations for nominal size range 尺寸分段的允许偏差(mm)							
Designation 等级	Description 属性	0.5~3	>3~6	>6~30	>30~120	>120~400	>400~1000	>1000~2000	>2000~4000
c	Coarse 粗糙	±0.2	±0.3	±0.5	±0.8	±1.2	±2	±3	±4
v	Very coarse 最粗糙	—	±0.5	±1	±1.5	±2.5	±4	±6	±6

Permissible Deviations for Broken Edges
(External Radii and Chamfer Heights)   Tab. 2-8
倒圆半径与倒角高度尺寸极限偏差数值   表2-8

Tolerance grade 公差等级		Permissible deviations for nominal size range 尺寸分段的允许偏差(mm)			
Designation 等级	Description 属性	0.5~3	>3~6	>6~30	>30
f	Fine 精密	±0.2	±0.5	±1	±2
m	Medium 中等	±0.2	±0.5	±1	±2
c	Coarse 粗糙	±0.4	±1	±2	±4
v	Very coarse 最粗糙	±0.4	±1	±2	±4

The value of the limit deviation of a hole or a shaft adopts the tolerance zone with symmetrical distribution. When general tolerance is adopted, nominal size shall be marked on the drawing without limit deviation, and standard number and tolerance grade code shall be used in the technical requirements of the drawing or relevant technical documents. For example, when the precision grade f is chosen, it is expressed as GB/T 1804-f.

无论是孔或者轴,其极限偏差的取值都采用了对称分布的公差带。当采用未注公差时,在图样上只标注公称尺寸,不标注极限偏差,在图样的技术要求或者有关技术文件中,用标准号和公差等级代号进行表示。例如:当选用精密级 f 时,则表示为 GB/T 1804-f。

2.3 Selection of Tolerances and Fits 公差与配合的选用

It is a very important work in mechanical designing to select tolerances and fits. It is the design of size precision under the condition that the nominal size has been determined. Its content includes three aspects: fit system, tolerance grade and fit type. Selection of basic system of fit, standard tolerance grades and fits are reasonable or not, not only influence the product performance and quality, but also affects the economic benefits. Therefore, reasonable selection of limits and fits can not only promote the interchangeability production, but also be conducive to improving the performance and the quality of products, and reduce the production cost. The principle of selection is to obtain the best technical and economic benefits on the premise of satisfying the use requirements. The methods of selection include calculation, experiment and analogy.

公差与配合的选择是机械设计与制造中的一个重要环节。它是在公称尺寸已经确定的情况下进行的尺寸精度设计,其内容包括配合制、公差等级和配合种类3个方面。公差与配合的选择是否恰当,对产品的性能、质量、互换性及经济性有着重要的影响。选择的原则是在满足使用要求的前提下能够获得最佳的技术经济效益。选择的方法有计算法、试验法和类比法。

2.3.1 Selection of Basic System of Fit 配合制的选择

The machinery structure, the machining and assembly processing, and the economy should be considered when selecting basic system of fits.

选择配合制时,应该从结构、工艺性和经济性几个方面综合分析考虑。

2.3.1.1 Hole-Basis System of Fit is Preferred 优先选用基孔制

Applying the hole-basis system can reduce the numbers and specifications of specified value cutting and measuring tools. Because a hole usually uses the setting tool (such as drill, reamer, broach etc.) to cut, and the smooth limit gauge to inspect, one tolerance zone should have one setting tool and limit gauge. Then the more hole tolerance zones are used, the more cutting tools are needed.

选用基孔制可以减少孔用定值刀具和量具等的数目。加工孔的刀具多是定值刀具,一个公差带就需要对应一个加工刀具,同时也需要一个对应的量具,改变了孔的尺寸,则会增加刀具和量具的数目。

2.3.1.2 Shaft-Basis System of Fit is Used for Some Special Cases 选用基轴制的情况

Shafts directly use cold drawn steel with certain accuracy. The shaft need not to cut, so it is economy that in these cases shaft-basis system fits is selected.

直接使用有一定公差等级而不再进行机械加工的冷拔钢材做轴时,轴不再进行机械加工,选择基轴制时经济性能最好。

According to the structure requirements, sometimes a shaft is mounted with several holes and the fit properties are different. In this case the shaft-basis system of fit should be used. In Fig. 2-14, the piston and the link rod in combustion engine mechanism, the fit is tight between piston pin and two holes in piston (transition fits), and the fit is loose between of three holes (the two pin holes in the piston and on hole in link rod) is same tolerance zone (H6), then to meet the two kinds of fits properties with two different tolerance zone in the pin (h5 in the middle and m5 in the both side), therefore, the pin is a stepped shaft. However, this will make the pin is not conducive machining, moreover, the hole in link rod will be scratched during assembly. Conversely, the shaft-basis system of fit is used, which the piston pin is used one kind of tolerance zone (h5) and

Chapter 2 Size Tolerances and Fits 尺寸公差与配合

the pin is plain shaft, so that machining and assembly of piston pin are easy.

根据结构上的需要，在同一公称尺寸的轴上装配有不同配合要求的几个孔件时，应采用基轴制。如图 2-14 所示是发动机的活塞销与连杆套孔和活塞孔之间的配合。根据工作需要及装配性，活塞销与活塞孔采用过渡配合，三个孔（活塞上的两个消控和连杆上的孔）之间的配合松，都具有相同的公差带(H6)。而活塞销与连杆套孔采用间隙配合。若采用基孔制配合，活塞销将做成阶梯状。采用基轴制配合，销可做成光轴。这种选择不仅有利于轴的加工，并且能够保证它们在装配中的配合质量。

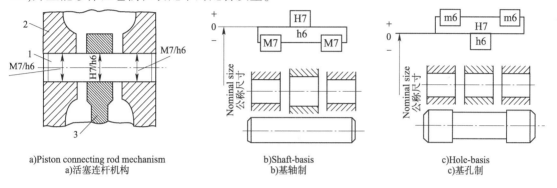

a) Piston connecting rod mechanism
a) 活塞连杆机构

b) Shaft-basis
b) 基轴制

c) Hole-basis
c) 基孔制

Fig. 2-14 Selection of Basic System of Fit(1)
图 2-14 配合制的选择(1)

2.3.1.3 Select Basic Fits System According to Standard Workpiece 与标准件配合

If mating with standard parts, standard parts should be taken as datum parts to determine whether basis hole system or basis shaft system is adopted. The shaft-basis system of fit is adopted between the housing hole and the external ring of the rolling bearing, however, the holebasis system of fit is adopted between a shaft and the internal ring of rolling bearing, because rolling bearing is a standard part.

若与标准件配合，应以标准件为基准件，来确定采用基孔制还是基轴制。例如，滚动轴承外圈与箱体孔的配合应该采用基轴制，滚动轴承内圈与轴的配合应采用基孔制。

2.3.1.4 Non-Datum System of Fit 非基准制的配合

The non-datum fit refers to the fit of two mating parts without the reference hole H or the reference shaft h. For a hole mating with several shafts or a shaft mating with several holes, the fit requirements are not the same, then some fits are non-datum fits, as shown in Fig. 2-15.

非基准制的配合是指相配合的两个零件既无基准孔 H 又无基准轴 h 的配合。当一个孔与几个轴相配合或一个轴与几个孔相配合，其配合要求各不相同时，则有的配合要出现非基准制的配合，如图 2-15 所示。

It is equipped with rolling bearing and bearing end cover in the box body hole. As the rolling bearing is a

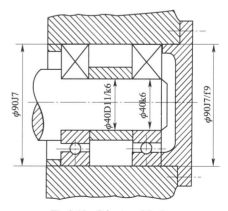

Fig. 2-15 Selection of Basic System of Fit(2)
图 2-15 配合制的选择(2)

standard part, its fit with the box body is based on shaft system, and the tolerance grade of the box body hole is code J7. In this case, if the fit between the bearing end cover and the box body hole should also adhere to the basis shaft system, the fit is J/h, which is a transition fit. However, because the bearing end cover needs to be removed frequently, it is obvious that the fit is too tight, and the clearance fit should be chosen. The bearing end cover tolerance grade cannot be h, only non-datum shaft tolerance grade can be selected. Considering the performance requirements of bearing end cover and economic efficiency of machining, tolerance grade 9 is adopted. Finally, J7/f9 is selected as the fit between bearing end cover and the box hole.

在箱体孔中装配有滚动轴承和轴承端盖,由于滚动轴承是标准件,它与箱体的配合是基轴制配合,箱体孔的公差带代号为J7,这时如果轴承端盖与箱体孔的配合也要坚持基轴制,则配合为J/h,属于过渡配合。但由于轴承端盖需要经常拆卸,显然这种配合过于紧密,而应该选用间隙配合。轴承端盖公差带不能用h,只能选择非基准轴公差带,考虑到轴承端盖的性能要求和加工的经济性,采用公差等级9级,最后选择的轴承端盖与箱体孔之间的配合为J7/f9。

2.3.2 Selection of Standard Tolerance Grades 公差等级的选择

To select size tolerance grades, it is to handle the contradiction between the functional requirements and the machining costs, complexity of manufacturing process. If the tolerance grade is too low, although the manufacturing process is simple and the cost is low, however the accuracy of the product is not guaranteed. In other hand, if the tolerance grade is too high, manufacturing processes is too complex, and then the cost increases. Both are not conducive to the comprehensive economic benefit and product market competitiveness. So we should consider the two sides of the contradiction to select a reasonable size tolerance grade.

标准公差等级的选用是一项重要的,同时又比较困难的工作,因为公差等级的高低直接影响产品使用性能和加工的经济性。公差等级过低,产品质量得不到保证;公差等级过高,将使制造成本增加。事实上,公差反映机器零件的使用要求与制造工艺成本之间的矛盾,所以在选用公差等级时,应兼顾这两方面的要求,正确合理地选用公差等级。

Basic principle for selecting the grade of tolerance is to choose a tolerance grade as low as possible in the context of meeting the design requirements. The methods to select size tolerance grades are usually according to the information and experiments from productive practice. The following aspects should be considered during selecting size tolerance grades:

选用标准公差等级的原则是:在充分满足使用要求的前提下,考虑工艺的可能性,尽量选用精度较低的公差等级。公差等级的选择常用类比法,即参考生产实践中总结出来的经验资料,进行比较选择。选择时应考虑如下几方面:

According to the standard, when the nominal size is less than 500mm, the tolerance grade of common fit is generally adopted, that is, the holes of grade 6, 7 and 8 are fitted with the shafts of grade 5, 6 and 7 respectively.

标准规定,公称尺寸小于500mm时,一般采用常用配合的公差的等级,即6、7、8级孔分别与5、6、7级轴配合。

Chapter 2   Size Tolerances and Fits 尺寸公差与配合

Generally speaking, IT2 ~ IT5 is selected as the fit is very precise. In general, IT5 ~ IT11 is used. For non-fit size, take IT12 ~ IT18, i.e. the tolerance range of general tolerance of linear size.

一般来说,配合特别精密时取 IT2 ~ IT5;一般配合,取 IT5 ~ IT11;非配合尺寸取 IT12 ~ IT18,即线性尺寸未注公差的公差等级范围。

The tolerance grade of particular use can be determined according to Tab. 2-9. The application of common dimensional tolerance grades are shown in Tab. 2-10.

表 2-9 为特殊用途下公差等级的选用依据。表 2-10 为常用尺寸公差等级的应用。

Application Scope of Individual Tolerance Grades　　　　Tab. 2-9
特殊用途下公差等级的选用　　　　表 2-9

Tolerance grades 公差等级	Application 应用
IT01 ~ IT1	For production of gauge blocks 用于量具、量块的生产
IT1 ~ IT7	For production of gauge to measure parts which are grades from IT6 to IT16 用于 IT6 至 IT16 等级量规的生产
IT5 ~ IT12	For fits in precision and general engineering 适用于精密和一般工程
IT8 ~ IT14	For production of semi-products or material 用于生产半成品或材料
IT12 ~ IT18	For non-fit sizes 用于非装配场合

Application of Common Dimensional Tolerance Grades　　　　Tab. 2-10
常用尺寸公差等级的应用　　　　表 2-10

Tolerance grades 公差等级	Application 应用
IT5	Used for very small size tolerance and geometrical tolerance, fit properties. Usually used in important fits in machine tools, engine, instruments etc, such as housing holes fitted with rolling bearing which are grade 5, spindles, tailstocks and sleeves in machine tool, precision machinery and high speed shafts, precision screw shafts with rolling bearing which are grade 6 主要应用在配合公差、形状公差要求很小的地方。一般在机床、发动机、仪表等重要部位应用。例如:与 5 级滚动轴承配合的箱体孔;与 6 级滚动轴承配合的机床主轴、机床尾座与套筒、精密机械及高速机械中轴、精密丝杠轴等
IT6	Fit properties are unified. Used for holes or shafts fitted with grade 6 rolling bearings, the shafts fitted with gears, worms, couplings, belt pulleys, cams etc, the column of radial drilling machine, guiding workpieces of machine jigs, basic hole of gears which are grades 6, basic shaft of gears which are grade 7 or 8 配合性质能达到较高的均匀性。例如:与 6 级滚动轴承相配合的孔、轴;与齿轮、涡轮、联轴器、带轮、凸轮等连接的轴,机床丝杠轴,摇臂钻床立柱,机床夹具中导向件外径尺寸;6 级精度齿轮的基准孔,7、8 级精度齿轮基准轴

Continued 续上表

Tolerance grades 公差等级	Application 应用
IT7	The precision of IT7 is lower than IT6, and it used general mechanical feature, such as holes of coupling, belt pulley, cams, holes of chuck seat in machine tools, fixed grill bush and changeable drill bush, basic holes of gears which are grade 7 or 8; basic shafts of gears which are grade 9 or 10 7级精度比6级稍低,应用条件与6级基本相似,在一般机械制造中应用较为普遍。例如:联轴器、带轮、凸轮等孔;机床卡盘座孔,夹具中固定钻套、可换钻套;7、8级齿轮基准孔,9、10级齿轮基准轴
IT8	It is of middle precision, used for the width size of bearing seats, basic holes of gears which grade 9~12, basic shafts of gears which are grade 11 or 12 在机械制造中属于中等精度。例如:轴承座衬套沿宽度方向尺寸;9~12级齿轮基准孔,11、12级齿轮基准轴
IT9~IT10	Mainly used for external surfaces and holes of shaft bush, operation feature and shafts; belt pulleys and shafts of which the axle is hollow, keys and springs 主要用于机械制造中轴套外径与孔,操纵件与轴,带轮与轴,单键与花键
IT11~IT12	The precision is very low, and the clearance is very large after assembly, so it is suitable for the condition with no fit requirements, such as flanges, slide blocks and slide gears, the in process size during machining; fit workpieces by punching. The connection of wrench hole and wrench seat in machine tool manufacturing 配合精度很低,装配后可能产生很大间隙,适用于基本上没有配合要求的场合。例如:机床上法兰盘与止口,滑块与滑移齿轮,加工中工序间尺寸,冲压加工的配合件,机床制造中的扳手孔与扳手座的连接

2.3.3 Selection of Fits 配合种类的选择

The choice of fit type is to select the basic deviation code of non-datum parts based on the determination of the datum system, according to the permitted clearance or interference size and its variation range in use, and sometimes determine at the same time the tolerance grade of datum parts and non-datum parts.

配合种类的选择就是在确定了基准制的基础上,根据使用中允许间隙或过盈的大小及其变化范围,选定非基准件的基本偏差代号,有时同时确定基准件与非基准件的公差等级。

Clearance, transition or interference fit shall be determined according to specific use requirements. For example, if the hole and shaft have relative motion requirements, the clearance fit must be selected; interference, transition and even clearance fit should be determined according to different working conditions when there is no relative movement of the hole and shaft. After determining the fits types, the preferred fits should be selected as far as possible, followed by the common fits, and finally the general mating. If still cannot meet the requirements, you may choose other fits.

确定间隙、过渡或过盈配合应根据具体的使用要求。例如：孔、轴有相对运动要求时，必须选择间隙配合；当孔、轴无相对运动时，应根据具体工作条件的不同确定过盈、过渡甚至间隙配合。确定配合种类后，应尽可能地选择优先配合，其次是常用配合，最后是一般配合。如果仍不能满足要求，可以选择其他配合。

There are three methods to select the basic deviation: calculation, experiment and analogy.

选定基本偏差的方法有3种，计算法、试验法和类比法。

The calculation method is to calculate the required clearance or interference size according to the theoretical formulas to select the fit method. For example, according to the theory of liquid lubrication, the minimum clearance required to guarantee the state of liquid friction is calculated; according to the elast oplasticity deformation theory, the minimum interference and the maximum interference that can ensure the transfer of a certain load are calculated. Because there are many factors affecting the clearance and interference, theoretical calculation gets only approximate results, so the actual choice needs to be determined through experiments. In general, it is rare to use the calculation method.

计算法就是根据理论公式，计算出满足使用要求的间隙或过盈的大小，来选定配合的方法。例如：根据液体润滑理论，计算保证液体摩擦状态所需要的最小间隙；根据弹性变形理论，计算出能够保证传递一定负载所需要的最小过盈和不使零件破坏的最大过盈。由于影响间隙和过盈的因素很多，理论计算得到的只是近似结果，所以实际使用的时候还需要通过试验来确定。一般情况下，很少用计算法。

The experiment method is to determine the clearance or interference range satisfying the working performance of the product by the experiment. This method is mainly used in the situation where these values have great influence on the product performance and people lack experience. The experiment method is more reliable, but the cycle is long and the cost is high, so it is only used for very important occasions.

试验法就是用试验的方法确定满足产品工作性能的间隙或过盈范围。该方法主要用于这些参数对产品性能影响大而又缺乏经验的场合。试验法比较可靠，但是周期长、成本高，所以只应用于非常重要的场合。

The analogy method is to refer to the same type of machine or mechanism through the production practice to verify the actual situation, and then combined with the design of the product to determine the use of requirements and application conditions. This method is the most widely used. To select the fit by the analogy method, we need to master the properties and applications of all kinds of fits first, especially to be more familiar with the common and priority fits stipulated by the national standards. Tab.2-11 lists the properties and applications of common basis hole fits in common size segments.

类比法就是参照同类型机器或机构中经过生产实践验证的配合实际情况，再结合所设计产品的使用要求和应用条件来确定配合类型。该方法应用最为广泛。用类比法选择配合，首先需要掌握各种配合的特征和应用场合，尤其是对国家标准所规定的常用与优先配合要更为熟悉。表2-11列出了常见尺寸段中基孔制常用配合的特征和应用场合。

Properties and Applications of Common Fits
常用尺寸基孔制常用配合的特征与应用

Tab. 2-11
表 2-11

Fit type 配合类别	Fit properties 配合特征	Symbols for fit 配合代号	Applications 应用场合
Clearance fit 间隙配合	Greatest clearance 特大间隙	$\frac{H11}{a11}\ \frac{H11}{b11}\ \frac{H12}{b12}$	Used for high temperature or great clearance fit needed 用于高温或工作要求大间隙的配合
	Greater clearance 很大间隙	$\frac{H11}{c11}\ \frac{H11}{d11}$	Used for very poor work condition, deformed under force, or great clearance need for assembly and high temperature work condition 用于工作条件较差、受力变形或为了便于装配而需要大间隙的配合和高温工作的配合
	Great clearance 较大间隙	$\frac{H9}{c9}\ \frac{H10}{c10}\ \frac{H8}{d8}\ \frac{H9}{d9}\ \frac{H10}{d10}\ \frac{H8}{e7}\ \frac{H8}{e8}\ \frac{H9}{e9}$	Used for sliding bearing under high speed and heavy load, for sliding bearing of big diameter, and for large span or multiple points supporting 用于高速重载的滑动轴承或大直径的滑动轴承,也可用于大跨距或多支点支撑的配合
	Common clearance 一般间隙	$\frac{H6}{f5}\ \frac{H7}{f6}\ \frac{H8}{f7}\ \frac{H8}{f8}\ \frac{H9}{f9}$	Used for running fit under common velocity, and widely used for sliding bearing under common lubrication, when temperature has little affection 用于一般转速的间隙配合。温度影响不大时,广泛应用于普通润滑油润滑的支撑处
	Smaller clearance 较小间隙	$\frac{H6}{g5}\ \frac{H7}{g6}\ \frac{H8}{g7}$	Used for precision sliding fit, or intermittence rotating fit worked at lowly speed 用于精密滑动零件或缓慢回转零件的配合部位
	Minimum clearance is 0 零间隙	$\frac{H6}{h5}\ \frac{H7}{h6}\ \frac{H8}{h7}\ \frac{H8}{h8}\ \frac{H9}{h9}\ \frac{H10}{h10}\ \frac{H11}{h11}\ \frac{H12}{h12}$	Used for common alignment fit, low moving fit or swing fit 用于不同精度要求的一般定位件的配合和缓慢移动和摆动的配合
Transition fit 过渡配合	Slightly clearance entirely 绝大部分有微小间隙	$\frac{H6}{js5}\ \frac{H7}{js6}\ \frac{H8}{js7}$	Used for concentrical fit with which easily mount and dismount or fit which transmits static load after fastening pieces added 用于易于装拆的定位配合或加紧固件后可传递一定静载荷的配合
	Slightly clearance mostly 大部分有微小间隙	$\frac{H6}{k5}\ \frac{H7}{k6}\ \frac{H8}{k7}$	Used for concentrical fit with a little vibration. Load can be transmitted after fastening pieces added, and easily mount and dismount; usually mounted by mallet 用于稍有振动的定位配合。加紧固件可传递一定载荷。拆装方便,可用木槌敲入
	Slightly interference mostly 大部分有微小过盈	$\frac{H6}{m5}\ \frac{H7}{m6}\ \frac{H8}{m7}$	Used for high precision concentrical and shake proof fits. Large load can be transmitted after fastening pieces added. Usually mounted by copper hammers 用于定位精度较高且能抗振的定位配合。加键可传递较大载荷。可用铜锤敲入或小压力压入

Chapter 2 Size Tolerances and Fits 尺寸公差与配合

Continued 续上表

Fit type 配合类别	Fit properties 配合特征	Symbols for fit 配合代号	Applications 应用场合
Transition fit 过渡配合	Slightly interference entirely 绝大部分有微小过盈	$\dfrac{H7}{n6}$ $\dfrac{H8}{n7}$	Used for extra precision concentrical fit or tight fit; heavy load or impacted load can be transmitted after fastening piece added such as key; dismounted only at overhaul 用于精确定位或紧密组合件的配合。加键能传递大力矩或冲击性载荷。只在大修时拆卸
	Slightly interference entirely 绝大部分有较小过盈	$\dfrac{H8}{p7}$	Heavy load can be transmitted after key added, and vibration and impact load can be sustained. Never dismounted 加键后能传递很大力矩,用于承受振动和冲击的配合。装配后不再拆卸
Interference fit 过盈配合	Slightly interference 轻型	$\dfrac{H6}{n5}$ $\dfrac{H6}{p5}$ $\dfrac{H7}{p6}$ $\dfrac{H6}{r5}$ $\dfrac{H7}{r6}$ $\dfrac{H8}{r7}$	Used for exact concentrical fit. Usually can bot transmit large load, if load need to transmit, fastening piece is still needed 用于精确的定位配合,一般不能靠过盈传递力矩。要传递力矩需要加紧固件
	Common interference 中型	$\dfrac{H6}{s5}$ $\dfrac{H7}{s6}$ $\dfrac{H8}{s7}$ $\dfrac{H6}{t5}$ $\dfrac{H7}{t6}$ $\dfrac{H8}{t7}$	Small load can be transmitted without fastening piece. Large load can be transmitted after fastening piece added 不需要加紧固件就可以传递较小力矩和进给力。加紧固件后可承受较大载荷或动载荷
	Great interference 重型	$\dfrac{H7}{u6}$ $\dfrac{H8}{u7}$ $\dfrac{H7}{v6}$	Large load can be transmitted without fastening piece and the material strength should be very high 不需要加紧固件就可以传递和承受大的力矩和动载荷。要求零件材料有高强度
	Greatest interference 特重型	$\dfrac{H7}{x6}$ $\dfrac{H7}{y6}$ $\dfrac{H7}{z6}$	Extra load can be transmitted, however, experiments needed before these fits used 能传递和承受很大力矩和动载荷,须经试验后方可应用

Exercises 2 习题 2

2-1 What are the meanings of nominal size, limit size, limit deviation and dimensional tolerance? How do they relate to each other? How is it represented on the tolerance zone diagram?

2-1 公称尺寸、极限尺寸、极限偏差和尺寸公差的含义是什么?它们之间的相互关系是什么?在公差带图上如何表示?

2-2 What is the standard tolerance factor? What is the standard tolerance?

2-2 什么是标准公差因子？什么是标准公差？

2-3 How to explain the deviations and the fundamental deviations? Why are fundamental deviations required?

2-3 如何解释偏差和基本偏差？为什么要规定基本偏差？

2-4 What is fit? What is the fit system? Why do we need to have a fit system?

2-4 什么是配合？什么是配合制？为什么要规定配合制？

2-5 The nominal size of the hole $D = 50$ mm, the upper limit size $D_{max} = 50.087$ mm, and the lower limit size $D_{min} = 50.025$ mm. The upper limit deviation ES, the lower limit deviation EI and tolerance T_D of the hole shall be required, and the tolerance zone diagram shall be drawn.

2-5 孔的公称尺寸 $D = 50$ mm，上极限尺寸 $D_{max} = 50.087$ mm，下极限尺寸 $D_{min} = 50.025$ mm，求孔的上极限偏差 ES、下极限偏差 EI 及公差 T_D，并画出公差带图。

2-6 The sizes of the hole that fit with each other are $\phi 15^{+0.027}_{0}$, and the sizes of the shaft are $\phi 15^{-0.016}_{-0.034}$. Try to calculate the ultimate size, ultimate deviation, dimensional tolerance, ultimate clearance (or interference), average clearance (or interference) and fit tolerance respectively, and draw the dimensional tolerance zone diagram and fit tolerance zone diagram.

2-6 相互配合的孔的尺寸为 $\phi 15^{+0.027}_{0}$，轴的尺寸为 $\phi 15^{-0.016}_{-0.034}$，试分别计算其极限尺寸、极限偏差、尺寸公差、极限间隙(或过盈)、平均间隙(或过盈)和配合公差，并画出尺寸公差带图与配合公差带图。

2-7 Known $\phi 30 \text{N7} \left({}^{-0.007}_{-0.028} \right)$ and $\phi 30 \text{t6} \left({}^{+0.054}_{+0.041} \right)$. Calculate the limit deviations of $\phi 30 \dfrac{\text{H7}}{\text{n6}}$ and $\phi 30 \dfrac{\text{T7}}{\text{h6}}$, and then draw the dimensional tolerance zone diagram.

2-7 已知 $\phi 30 \text{N7} \left({}^{-0.007}_{-0.028} \right)$ 和 $\phi 30 \text{t6} \left({}^{+0.054}_{+0.041} \right)$。计算 $\phi 30 \dfrac{\text{H7}}{\text{n6}}$ 与 $\phi 30 \dfrac{\text{H7}}{\text{h6}}$ 的极限偏差，并画出尺寸公差带图。

Chapter 3 Geometrical Tolerances
几何公差

3.1 Basic Concepts 基本概念

3.1.1 Definition of Geometric Tolerances 几何公差定义

Mechanical parts are manufactured through design, processing and other processes. The workpieces are consisted by the features with perfect shape, orientation and location in the engineer drawings during designing, however, during manufacturing, the shape, orientation, location of a geometrical feature of a workpiece will never be ideal as designed, i.e. there will be form errors, orientation errors and location errors except size deviations. All these errors are called geometry errors. Geometric tolerance is given in design, which is used to control the geometric error produced by machining.

机械零件是通过设计、加工等过程制造出来的。在设计阶段，图样上给出的零件都是没有误差的几何体，构成这些几何体的点、线、面都具有理想几何特征。加工后零件的实际几何体和理想几何体之间存在差异。这些表现在零件几何元素的形状、方向、位置上的偏差，分别称为形状误差、方向误差和位置误差，统称为几何误差。几何公差就是在设计时给出的，用于控制加工产生的几何误差。

3.1.1.1 Cause of Geometrical Errors 几何误差产生的原因

There are lots of causes for geometrical errors of workpiece during manufacturing. First, there are errors of the process system, which consisted of machining tools, fixtures, cutting tools and workpiece. Second, the workpiece will be deformed by force and thermo during manufacturing. Third, there are vibrations and the cutting tools will be worn. So it is inevitable that there are errors of the feature which be cut.

在制造过程中，造成工件几何误差的原因有很多。例如：加工刀具、夹具、刀具和工件组成的工艺系统误差，在制造过程中工件受力、受热变形，工作台振动，刀具的磨损等。

3.1.1.2 Effects on Performance of the Geometrical Errors 几何误差对零件性能的影响

The effects of the geometrical errors on performance include the following aspects: Effect on assemblability, such as the location errors of holes will cause difficult to insert the bolts; effect on fit properties, such as the form errors of the contact surface of the hole and the shaft will cause the gap distribution is non-uniform in clearance fit, then the movement is lopsided if there is motion relatively; in interference fit, the interference value is non-uniform, and then the strength of connection will be decreased; effect on work accuracy, such as the straightness error of lathe bed guide rail will influence the movement accuracy of the saddle, and the form error and location error of two support journal of the lathe spindle will influence the rotary accuracy of the lathe spindle; effect on other function, such as the sealing will be influenced if there are geometrical errors in hydraulic system, and the contact area will be decreased if there are form errors on the contact surfaces, then the contact stiffness and the capacity of bearing load.

几何误差对零件性能的影响包括以下几个方面：可装配性的影响，如孔的位置误差会导致螺栓难以插入；配合性能的影响，如孔与轴接触面的形状误差，会导致间隙配合中的间隙分布不均匀，如果有相对运动，则运动是不平衡的；过盈配合时，过盈量不均匀，会降低连接强度；工作精度的影响，如机床导轨的直线度误差会影响运动精度，车床主轴两支承轴颈的形状误差和位置误差会影响机床主轴的回转精度；液压系统中存在几何误差会影响密封等其他功能，接触面存在形状误差会使接触面积减小，接触刚度和承载能力降低。

3.1.2　Geometric Features　几何要素

The research object of geometric tolerance is the geometric features of mechanical parts. Geometric features are the general designation of points, lines and surfaces that constitute the geometric features of parts. In order to facilitate the study of geometric tolerances and geometric errors, these geometric features can be classified in different ways.

几何公差的研究对象是机械零件的几何要素。几何要素是构成零件几何特征的点、线、面的统称。为了便于研究几何公差和几何误差，这些几何要素可以按不同角度进行分类。

According to the structural characteristics, they are divided into integral features and derived features. The surface and the lines on the surface of a part are called integral features, such as cylinder surface, two end surface and plain line on a cylinder surface. The points, lines and surfaces that make up the symmetry center of features are called derived features, such as the central shaft of a cylinder and the center of a sphere.

几何要素按结构特征分为组成要素和导出要素。构成零件的面及面上的线称为组成要素，例如圆柱体的圆柱面、两端面、圆柱面上素线。组成要素对称中心的点、线、面各要素称为导出要素，例如圆柱体的中心轴线、球体的球心。

According to the existing state, they are divided into nominal feature and actual feature. In the drawing, only geometric features are called nominal features. Nominal features have no error, and the features represented by mechanical drawings are nominal features. The actually existing features of parts are called actual features, which are usually replaced by measured features.

几何要素按存在状态分为公称要素和实际要素。在图样中只具有几何意义的要素称为公称要素，公称要素没有任何误差，机械图样所表示的要素均为公称要素。零件实际存在的要素称为实际要素，通常用测量得到的要素替代。

According to their positions, they are divided into measured features and datum features. The feature that gives the geometric tolerance requirements on the drawing is called the measured feature. The features used to determine the direction and position of the measured features are called datum features.

Chapter 3 Geometrical Tolerances 几何公差

几何要素按所处地位分为被测要素和基准要素。图样上给出了几何公差要求的要素称为被测要素。用来确定被测要素方向和位置的要素称为基准要素。

According to the functional requirements, they are divided into individual features and related features. A individual feature is a feature that only requires the form tolerance of the feature itself. The feature which has functional requirements relative to datum feature and gives orientation, position and run-out tolerance is called related feature.

几何要素按功能要求分为单一要素和关联要素。仅对要素本身提出形状公差要求的要素称为单一要素。相对于基准要素有功能要求而给出方向、位置和跳动公差的要素称为关联要素。

3.1.3 Symbols of Geometric Tolerances 几何公差符号

There are 14 geometrical tolerance characteristics in *Geometrical product specifications (GPS)—Geometrical tolerancing—Tolerances of form, orientation, location and run-out* (international standard ISO 1101 and national standard GB 1182—2018), shown in Tab. 3-1.

根据国际标准(ISO 1101)和国家标准《产品几何技术规范(GPS) 几何公差 形状、方向、位置和跳动公差标注》(GB 1182—2018)的规定,几何公差的14个特征项目分为形状、方向、位置和跳动公差4大类,它们的名称和符号见表3-1。

Features and Symbols of Geometric Tolerances Tab. 3-1
几何公差特征符号 表3-1

Features and Tolerances 公差特征与类型		Toleranced characteristics 几何特点	Symbols 符号
Individual features 单一要素	Form tolerances 形状公差	Straightness 直线度	─
		Flatness 平面度	▱
		Circularity 圆度	○
		Cylindricity 圆柱度	⌭
Individual or related features 单一或关联要素		Profile of a line 线轮廓度	⌒
		Profile of a surface 面轮廓度	⌓
Related features 关联要素	Orientation tolerances 方向公差	Parallelism 平行度	∥
		Perpendicularity 垂直度	⊥
		Angularity 倾斜度	∠
	Location tolerances 位置公差	Position 位置度	⌖
		Concentricity and coaxiality 同轴度	◎
		Symmetry 对称度	=
	Run-out tolerances 跳动公差	Circular run-out 圆跳动	↗
		Total run-out 全跳动	⌰

The geometrical tolerances are divided into four categories, i. e. form tolerances, orientation tolerances, location tolerances and run-out tolerances. There are no datum for form tolerances, however, there should be datum for orientation, location and run-out tolerances. Profile of any line and profile of any surface can belong to form, orientation and location tolerance according to its function requirements. Form tolerances also include profile of line and profile of surface, where straightness and circularity are used to limit form error of straight line and form error of circle respectively, and flatness and cylindricity are used to limit form error of plane and form error of cylindrical surface. Orientation tolerances are used to limit angular error between different features. Parallelism and perpendicularity are used to limit special angle. Parallelism is used to limit angular error where toleranced feature is 0 degree to datum features and perpendicularity is used to limit angular error where toleranced features is 90 degree to datum features. Location tolerances are used to limit location error of the toleranced feature with datum features. Coaxiality and symmetry are used to limit two special location errors. Run-out is defined by measurement method, and it is a comprehensive tolerance which can limit form, orientation and location errors.

几何公差分为形状公差、方向公差、位置公差和跳动公差4种类型。形状公差没有基准，但是方向公差、位置公差和跳动公差有基准。根据其功能要求，线轮廓度和面轮廓度可属于形状公差，也可以属于方向公差。形状公差包括线轮廓度公差和面轮廓度公差，其中直线度和圆度分别用于限制直线形状误差和圆度形状误差，平面度和圆柱度用于限制平面形状误差和圆柱面形状误差。方向公差用于限制不同特征之间的角度误差。平行度和垂直度用于限制特殊角度，平行度用于限制被测要素与基准要素成0°误差，垂直度用于限制被测要素与基准要素成90°误差。位置公差用于限制具有基准特性的被测要素的位置误差。同轴度和对称度用来限制两种特殊的定位误差。跳动量是用测量方法定义的，它是一种综合公差，可以限制形状误差、方向误差和位置误差。

3.1.4 Marking of Geometric Tolerances 几何公差标注

Geometric tolerance is marked in the form of tolerance frame on the drawing, as shown in Fig. 3-1. A tolerance frame is divided into two or three rectangular frames. When the measured feature is form tolerance, the tolerance frame has only the first two sections. When the measured feature is orientation, position or run-out tolerance, the tolerance frame is composed of three sections, the third, optional, datum section may consist of one to three compartments. The leader with an arrow should point to the corresponding measured feature.

几何公差在图样上用公差框格的形式标注，如图3-1所示。公差框格包括2部分或3部分矩形框格。当被测要素是形状公差时，公差框格只有前面2部分。当被测要素是方向、位置或跳动公差时，公差框格是3部分，第3部分是基准部分可包含1～3格。带箭头的指引线应指向相应的被测要素。

If the geometrical tolerance is specified to integral feature, the leader line arrow should be placed on the outline of the feature or an extension of the feature, see Fig. 3-2.

当被测要素为组成要素时，指引线的箭头应放置于要素的轮廓线或其延长线上，并与尺寸线明显错开，如图3-2所示。

Chapter 3　Geometrical Tolerances 几何公差

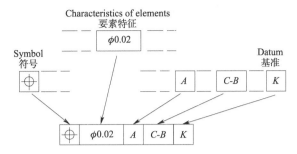

Fig. 3-1　Geometrical Tolerance Frame
图 3-1　几何公差框格

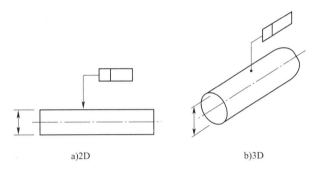

Fig. 3-2　Marking of the Integral Feature
图 3-2　组成要素标注

When the toleranced feature is a derived feature, it is indicated by the leader line starting either end of the tolerance frame terminated by an arrow on the extension of the size line of a feature of size, see Fig. 3-3.

当被测要素是导出要素时,指引线的箭头应该与该要素的尺寸线对齐,如图 3-3 所示。指引线原则上只能从公差框格的一端引出一条,可以曲折,但一般不得多于两次。

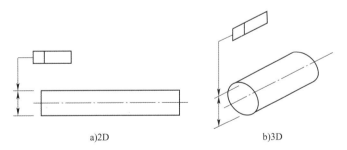

Fig. 3-3　Marking of Derived Feature
图 3-3　导出要素标注

The datum symbol is composed of capital letters with small squares connected with small black triangles by thin solid lines, as shown in Fig. 3-4. The letter indicating the datum is called the datum letter. Generally, I, O, Q and X are not recommended. Regardless of the orientation of the datum symbol on the drawing, the letters in the small square of the datum symbol shall be written horizontally. The single datum is represented by one letter; the common datum is represented by two letters separated by horizontal lines; the datum system is represented by two or three letters, arranged from left to right according to the sequence of datum, Ⅰ, Ⅱ and Ⅲ datum. When the datum feature is a contour line or surface, the small black isosceles triangle of the datum symbol is close to the contour line or its extension line of the datum feature, and the datum line and the size line of the contour are obviously staggered, as shown in Fig. 3-5. When the datum feature is a shaft or a central plane or determined by a feature with sizes, the datum line of the datum symbol is aligned with the size line, as shown in Fig. 3-6.

基准符号由带小方格的大写英文字母用细实线与小黑色三角形相连而组成,如图3-4所示。表示基准的字母称为基准字母,一般不建议采用I、O、Q、X这4个英文字母。无论基准符号在图面上的方向如何,基准符号小方格中的字母都应水平书写。单一基准由1个字母表示;公共基准采用由横线隔开的2个字母表示;基准体系由2个或3个字母表示,按基准的先后顺序从左至右排列,分别为第Ⅰ基准、第Ⅱ基准和第Ⅲ基准。当基准要素是轮廓线或表面时,基准符号的小黑色等腰三角形靠近基准要素的轮廓线或其延长线,且基准连线与轮廓的尺寸线明显错开,如图3-5所示。当基准要素是轴线或中心平面或由带尺寸的要素确定时,则基准符号的基准连线与尺寸线对齐,如图3-6所示。

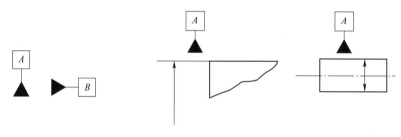

Fig. 3-4　Symbol of Datum　　　　Fig. 3-5　Datum Marking Method of Integral Features
图3-4　基准符号　　　　　　　　图3-5　组成要素的基准标注

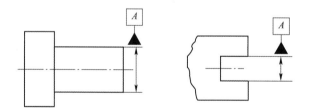

Fig. 3-6　Datum Marking Method of Derived Features
图3-6　导出要素的基准标注

If tolerances of location, orientation or profile are prescribed for a feature or a group of features, the sizes determining the theoretically exact location, orientation or profile respectively are called theoretically exact sizes (TED). TED shall not be toleranced, and they are to be enclosed in a frame, see the examples in Fig. 3-7. TED also applies to the sizes determining the relative orientation of the datum of a system.

对于要素的位置度、方向度或轮廓度,其尺寸由不带公差的理论正确位置、轮廓或角度确定,这种尺寸称为理论正确尺寸。理论正确尺寸应围以框格。零件实际尺寸仅由公差框格中的位置度、方向度或轮廓度公差值来限定,如图3-7所示。

The meaning of extended tolerance zone is to extend the tolerance zone of measured features

Chapter 3　Geometrical Tolerances 几何公差

beyond the workpiece entity, and control the tolerance zone outside the workpiece, so as to ensure that the mating parts can be loaded smoothly when fitting with the part. The extended tolerance zone is represented by a symbol Ⓟ and its extension range is indicated, as shown in Fig. 3-8.

延伸公差带的含义是将被测要素的公差带延伸到工件实体之外，控制工件外部的公差带，以保证相配零件与该零件配合时能顺利装入。延伸公差带用符号Ⓟ表示，其延伸的范围被注出，如图 3-8 所示。

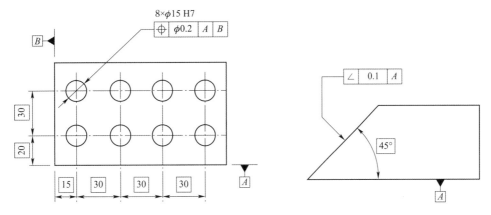

Fig. 3-7　Examples of TED Used
图 3-7　理论正确尺寸标注示例

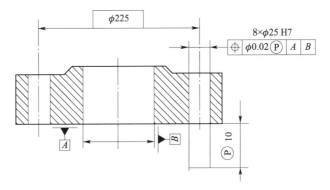

Fig. 3-8　Extended Tolerance Zone
图 3-8　延伸公差带

3.1.5　Geometric Tolerance Zone 几何公差带

The geometric tolerance zone is used to limit the variation area of the measured actual feature. It is a geometric figure. As long as the measured actual features are completely within the given tolerance zone, the geometric accuracy of the measured actual features meets the design requirements.

几何公差带用来限制被测实际要素变动的区域，它是一个几何图形。只要被测实际要素完全落在给定的公差带内，就表示被测实际要素的几何精度符合设计要求。

Geometric tolerance zone has four features: shape, size, direction and position. The shape of the geometric tolerance zone is determined by the ideal shape of the measured feature and the given tolerance feature. The size of the geometric tolerance zone is determined by the tolerance value, which refers to the width or diameter of the tolerance zone. The direction of geometric tolerance zone refers to the direction perpendicular to the extension direction of tolerance zone, usually the direction indicated by the leader arrow. The position of geometric tolerance zone can be fixed or floating. Once the position of the datum feature on the drawing is determined, the position of the tolerance zone will not change any more, it is called the position fixed of the tolerance zone. When the position of the tolerance zone can change with the actual size, it is called the position floating of the tolerance zone. For example: for cylindricity, the tolerance zone is coaxial and fixed with the datum shaft; for flatness, the floating degree varies with the actual position of the plane.

几何公差带具有形状、大小、方向和位置4个要素。几何公差带的形状由被测要素的理想形状和给定的公差特征所决定。几何公差带的大小由公差数值确定，指的是公差带的宽度或直径等。几何公差带的方向是指与公差带延伸方向相垂直的方向，通常为指引线箭头所指的方向。几何公差带的位置有固定和浮动两种。当图样上基准要素的位置一经确定，其公差带的位置不再变动，则称为公差带的位置固定。当公差带的位置可随实际尺寸的变化而变动，则称为公差带的位置浮动。例如：同轴度，其公差带与基准轴线共轴而且固定；平面度，随实际平面所处的位置不同而浮动。

3.2　Definition of Form Tolerances 形状公差

Form tolerance limits the deviation of a feature from its geometrical ideal line or surface form. Form tolerance is defined form tolerance zone. Form tolerance zone limits the permitted space for the actual single feature, i.e. the form is qualified when all point of the feature within the space. Form tolerance zone does not involve datum, has no clear direction and fixed position, and floats with the change of actual features.

形状公差限制了被测要素的几何特征与其几何理想线或曲面形状的偏差。形状公差的特征体现在形状公差带上。形状公差带限制实际被测要素单个特征的允许空间，即当所有特征点都在该空间内时，形状是合格的。形状公差带不涉及基准，没有明确的方向和固定的位置，随着实际要素的变动而浮动。

All form tolerances apply to single or individual features. Consequently, form tolerances are independent of all other features, so no datum applies to form tolerances. Form tolerance zone only limit the size of form error, the orientation and location of the form tolerance zone (except profile degree) is floating.

所有形状公差适用于单一要素（特征）。因此，形状公差独立于所有其他特征，形状公差没有基准。形状公差带只是限制形状误差的大小，除了轮廓度，形状公差带的方向和位置是浮动的。

In all the form tolerances, when different tolerance characteristics applied to same feature, the flatness can control the straightness error, the cylindricity can control the circularity error and the profile of any surface can control the error of profile of any line, because the later is part of the former. All the points are within the tolerance zone, and then the part of the points is also within the tolerance zone. If there is stricter requirement to the later tolerance, the tolerance value of later should be smaller than former tolerance. For Example, the flatness value is larger than the straightness value and the cylindricity value is larger than the circularity value.

在所有的形状公差中,对同一特征应用不同的公差特性时,平面度控制直线度误差,圆柱度控制圆度误差,任意曲面的轮廓度控制任意直线的轮廓度误差,因为后者是前者的一部分。整体要素点都在公差带内,则局部要素点也在公差带内。对同一被测要素提出两个几何公差要求,如果对后提出的公差有更严格的要求,可以指定后者的公差值应小于前者。例如:平面度值大于直线度值,圆柱度值大于圆度值。

3.2.1　Straightness Tolerances 直线度公差

Straightness tolerance is used to limit the shape error of the straight line in plane or space. The measured feature can be a integral feature or a derived feature.

直线度公差用于限制平面内或空间直线的形状误差。被测要素可以是组成要素或导出要素。

In the specific plane or the considered plane, the tolerance zone is limited by two parallel straight lines with a distance of t and in the specified plane and direction. The location of the tolerance zone is float. It is qualified during any extracted line on the upper surface parallel to the plane of projection is contained between two parallel straight lines 0.1mm apart, as shown in Fig. 3-9.

给定平面内的直线度,公差带为在给定平面内和给定方向上,间距等于公差值 t 的两个平行直线所限定的区域。公差带的位置是浮动的。如图 3-9 所示,在任一平行于图示投影面的平面内,上表面的实际投影线应限定在间距等于 0.1mm 的平行直线之间。

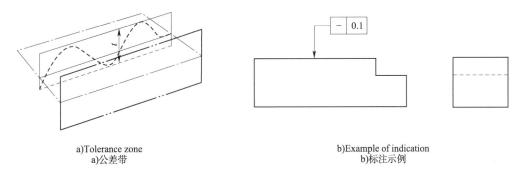

a) Tolerance zone　　　　　　　　　b) Example of indication
a) 公差带　　　　　　　　　　　　b) 标注示例

Fig. 3-9　Straightness Tolerances in the Specific Plane
图 3-9　给定平面内直线度公差

In the particular direction, the tolerance zone is limited by two parallel planes with a distance of t. The location of the tolerance zone is float. It is qualified during any extracted generating line on the cylindrical surface is contained between two parallel straight lines 0.1mm apart, as shown in Fig. 3-10.

在给定方向上,公差带为间距等于公差值 t 的两平行直线所限定的区域。公差带的位置是浮动的。如图 3-10 所示,实际轮廓线或素线应限定在间距等于 0.1mm 的两个平行直线之间。

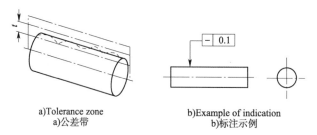

a)Tolerance zone
a)公差带
b)Example of indication
b)标注示例

Fig. 3-10 Straightness Tolerances in the Particular Direction
图 3-10 给定方向上直线度公差

In the random direction, the tolerance zone is limited by a cylinder of diameter t, if the tolerance value is preceded by the symbol ϕ. The location of the tolerance zone is float. It is qualified during the extracted median line of the cylinder to which the tolerance applies is contained within a cylindrical zone of diameter 0.08mm, as shown in Fig. 3-11.

在任意方向上,如果公差值前加了 ϕ,则公差带为直径等于公差值 t 的圆柱面所限定的区域。公差带位置是浮动的。如图 3-11 所示,外圆柱面的实际轴线应限定在直径等于 ϕ0.08mm 的圆柱面内。

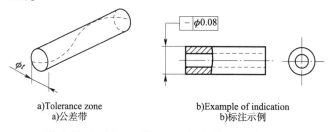

a)Tolerance zone
a)公差带
b)Example of indication
b)标注示例

Fig. 3-11 Straightness Tolerances in the Random Direction
图 3-11 任意方向上直线度公差

3.2.2 Flatness Tolerances 平面度公差

Flatness tolerance is used to limit the shape error of the measured actual plane. The feature to be measured can be a integral feature or a derived feature. The attribute and the shape of the nominal measured feature is a given surface, which belongs to the face feature.

平面度公差用以限制被测实际平面的形状误差。被测要素可以是组成要素或导出要素,其公称被测要素的属性和形状为明确给定的表面,属于面要素。

In the specific direction, the tolerance zone is limited by two parallel planes with a distance of t. It is qualified during the extracted surface is contained between two parallel planes 0.08mm apart, as shown in Fig. 3-12.

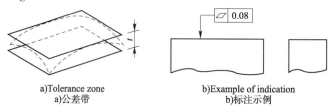

a)Tolerance zone
a)公差带
b)Example of indication
b)标注示例

Fig. 3-12 Flatness Tolerances in the Specific Direction
图 3-12 给定方向上平面度公差

平面度在给定方向上,公差带为间距等于公差值 t 的两平行平面所限定的区域。公差带位置是浮动的。如图 3-12 所示,实际表面应限定在间距等于 0.08mm 的两平行平面之间。

3.2.3　Circularity Tolerances 圆度公差

The circularity tolerance is used to limit the shape error of the profile in one direction of the revolving surface. The measured feature is a integral feature, and the measured feature is a given circumference or a group of circumferences. The circularity requirements of cylindrical feature can be used on the sections perpendicular to the shaft of the measured feature. Cylindrical requirements for spherical features can be used on sections containing spheres. The rotating surface of non-cylinder should be marked with directional feature.

圆度公差用以限制回转表面的某一方向截面轮廓的形状误差。被测要素是组成要素,且被测要素是明确给定的圆周线或一组圆周线。圆柱要素的圆度要求可用在被测要素轴线垂直的截面上;球形要素的圆度要求可用在包含球形的截面上;非圆柱体的回转表面应标注方向要素。

The tolerance zone, in the considered cross-section, is limited by two concentric circles with a difference in radii of t. The location of the tolerance zone is float. It is qualified during the extracted circumferential line, in any cross-section of the cylinder, and is contained between two co-planar concentric circles with a different in radii of 0.03mm. It is qualified during any extracted circumferential line, resulting in the intersection of the revolute surface and a section cone shall be contained in a conical zone limited by two circles 0.03mm apart, as shown in Fig. 3-13.

圆度公差带为在给定截面内,半径差等于公差值 t 的两同心圆所限定的区域。公差带位置是浮动的。如图 3-13 所示,在圆柱面的任一横截面内,实际圆周应限定在半径差等于 0.03mm 的两个共面同心圆之间;在圆锥面的任一横截面内,实际圆周应限定在半径差等于 0.03mm 的两个共面同心圆之间。

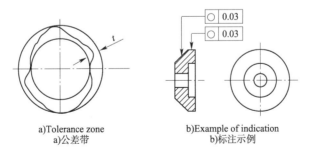

a)Tolerance zone　　　　b)Example of indication
a)公差带　　　　　　　　b)标注示例

Fig. 3-13　Circularity Tolerances
图 3-13　圆度公差

3.2.4　Cylindricity Tolerances 圆柱度公差

The cylindricity tolerance is used to limit the shape error of the measured actual cylindrical surface. The cylindricity tolerance is only the control requirement of cylindrical surface, and cannot be used for conical surface or other shaped surface. The cylindricity tolerance controls the shape errors of cylinder cross section and axial section, such as circularity, straightness of prime line, straightness of shaft, etc. Therefore, the cylindricity is the comprehensive control index for various shape errors of cylindrical surface. The cylindricity leader arrow is perpendicular to the profile surface.

圆柱度公差用以限制被测实际圆柱面的形状误差。圆柱度公差仅是对圆柱表面的控制要求,它不能用于圆锥表面或其他形状的表面。圆柱度公差同时控制了圆柱体横剖面和轴向剖面内各项形状误差,诸如圆度、素线直线度、轴线直线度误差等。因此,圆柱度是圆柱面各项形状误差的综合控制指标。圆柱度的指引线箭头垂直于轮廓表面。

The tolerance zone is limited by two coaxial cylinders with a difference in radii of t. The location of the tolerance zone is float. The extracted cylindrical surface is contained between two coaxial cylinders with a different in radii of 0.1mm, as shown in Fig. 3-14.

圆柱度公差带为半径差等于公差值 t 的两同轴圆柱面所限定的区域。公差带位置是浮动的。如图 3-14 所示,实际圆柱面应限定在半径差等于 0.1mm 的两同轴圆柱面之间。

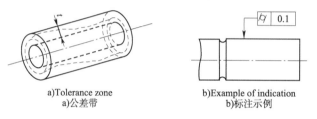

a)Tolerance zone　　　　　b)Example of indication
a)公差带　　　　　　　　b)标注示例

Fig. 3-14　Cylindricity Tolerances

图 3-14　圆柱度公差

3.3　Profile Tolerances 轮廓度公差

The profile tolerance includes line profile tolerance and surface profile tolerance. When there is no datum requirement, it is form tolerance; when there is datum requirement, it is orientation or location tolerance.

轮廓度公差包括线轮廓度公差和面轮廓度公差。无基准要求时为形状公差,有基准要求时为方向公差、位置公差。

3.3.1　Profile Tolerances of Any Line 线轮廓度公差

When there is no datum, the tolerance zone is limited by two lines enveloping circles of diameter t, the centers of which are situated on a line having the theoretically exact geometrical form. The location of the tolerance zone is float. It is qualified during the extracted profile line, in any section parallel to the projection, and is contained between two equidistant lines enveloping circles of diameter 0.04mm, the centers of which are situated on a line having the theoretically exact geometrical form, as shown in Fig. 3-15.

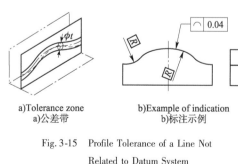

a)Tolerance zone　　　b)Example of indication
a)公差带　　　　　　b)标注示例

Fig. 3-15　Profile Tolerance of a Line Not Related to Datum System

图 3-15　无基准时的线轮廓度公差

没有基准时,公差带为直径等于公差值 t、圆心位于被测要素理论正确几何形状上的一系列圆的两包络线所限定的区域。公差带位置是浮动的。如图 3-15 所示,在任一平行于图示投影面的截面内,实际轮廓线应限定在直径等于 0.04mm、圆心位于被测要素理论正确几何形状上的一系列圆的两包络线之间。

When there is a datum, the tolerance zone is limited by two lines enveloping circles of diameter t, the centers of which are situated on a line having the theoretically exact geometrical form with respect to datum. The location of the tolerance zone is fixed. It is qualified, during in each section, parallel to the plane of projection or datum plane A, extracted profile line shall be contained between two equidistant lines enveloping circles of diameter 0.04mm, the centers of which are situated on a line having the theoretically exact geometrical form with respect to datum plane A and datum plane B, as shown in Fig. 3-16.

有基准时,公差带为直径等于公差值 t、圆心位于由基准确定的被测要素理论正确几何形状上的一系列圆的两包络线所限定的区域。公差带位置是固定的。如图 3-16 所示,在任一平行于图示投影面的截面内,实际轮廓线应限定在直径等于 0.04mm、圆心位于由基准 A 和基准 B 确定的被测要素理论正确几何形状上的一系列圆的两包络线之间。

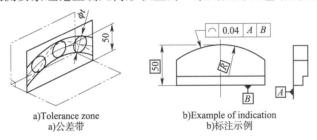

a) Tolerance zone
a) 公差带

b) Example of indication
b) 标注示例

Fig. 3-16　Profile Tolerance of a Line Related to Datum System
图 3-16　有基准时的线轮廓度公差

3.3.2　Profile Tolerances of Any Surface 面轮廓度公差

When there is no datum, the tolerance zone is limited by two surfaces enveloping spheres of diameter t, the centers of which are situated on a surface having the theoretically exact geometrical form. The location of the tolerance zone is float. It is qualified during the extracted surface is contained between two equidistant surfaces enveloping spheres of diameter 0.02mm, the centers of which are situated on a surface having the theoretically exact geometrical form, as shown in Fig. 3-17.

没有基准时,公差带为直径等于公差值 t、球心位于被测要素理论正确几何形状上的一系列圆球的两包络面所限定的区域。公差带位置是浮动的。如图 3-17 所示,实际轮廓面应限定在直径等于 0.02mm、球心位于被测要素理论正确几何形状上的一系列圆球的两包络面之间。

When there is a datum, the tolerance zone is limited by two surfaces enveloping spheres of diameter t, the centers of which are situated on a surface having the theoretically exact geometrical form with respect to datum. The location of the tolerance zone is fixed. It is qualified during the extracted surface shall be contained between two equidistant surfaces enveloping spheres of diameter 0.1mm, the centers of which are situated on a surface having the theoretically exact geometrical form with respect to datum A, as shown in Fig. 3-18.

有基准时,公差带为直径等于公差值 t、球心位于由基准确定的被测要素理论正确几何形状上的一系列圆球的两包络面所限定的区域。公差带位置是固定的。如图 3-18 所示,实际轮廓面应限定在直径等于 0.1mm、球心位于由基准平面 A 确定的被测要素理论正确几何形状上的一系列圆球的两包络面之间。

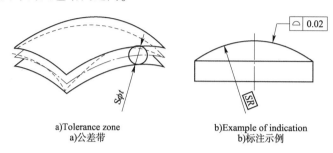

a)Tolerance zone　　　　　b)Example of indication
a)公差带　　　　　　　　b)标注示例

Fig. 3-17　Profile Tolerance of a Surface Not Related to Datum System

图 3-17　无基准时的面轮廓度公差

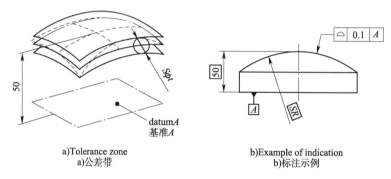

a)Tolerance zone　　　　　b)Example of indication
a)公差带　　　　　　　　b)标注示例

Fig. 3-18　Profile Tolerance of a Surface Related to Datum System

图 3-18　有基准时的面轮廓度公差

3.4　Orientation Tolerances　方向公差

The orientation tolerance is the allowable variation of the related actual features to the datum in the direction. The orientation tolerance zone has a clear direction relative to the datum and has the function of controlling the direction and shape of the measured feature at the same time.

方向公差是关联实际要素对基准在方向上的允许变动量。方向公差带相对基准具有明确的方向,具有同时控制被测要素方向和形状的功能。

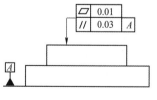

Fig. 3-19　Parallelism Tolerance and Flatness Tolerance are Given at the same time

图 3-19　同时给出方向公差和形状公差

On the premise of ensuring the functional requirements, the form tolerance is not given after the orientation tolerance is given. If there are further requirements for shape accuracy, form tolerance and orientation tolerance can be given at the same time, but the tolerance value of form tolerance should be less than that of orientation tolerance. The parallelism tolerance and flatness tolerance are 0.03mm and 0.01mm respectively, as shown in Fig. 3-19.

Chapter 3 Geometrical Tolerances 几何公差

在保证功能要求的前提下,对被测要素给出方向公差后,一般不再给出其形状公差。如果对形状精度有进一步要求时,可同时给出形状公差和方向公差,但是形状公差的公差值应该小于方向公差的公差值。如图 3-19 所示,对被测平面给出 0.03mm 的平行度公差和 0.01mm 的平面度公差。

3.4.1 Parallelism Tolerances 平行度公差

Parallelism tolerance of a plane related to a datum plane, the tolerance zone is limited by two parallel planes with a distance of t and parallel to the datum plane. The direction of tolerance zone is fixed. It is qualified during the extracted surface shall be contained between two parallel planes 0.01mm apart, which are parallel to the datum plane A, as shown in Fig. 3-20.

被测平面对基准平面 A 的平行度,其公差带为间距等于公差值 t 且平行于基准平面的两平行平面所选定的区域,公差带的方向是固定的。如图 3-20 所示,实际表面应限定在间距等于 0.01mm 且平行于基准平面的两平行平面之间。

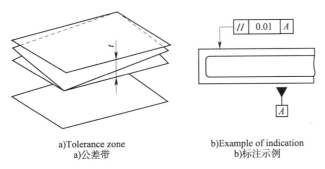

a)Tolerance zone a)公差带   b)Example of indication b)标注示例

Fig. 3-20 Parallelism Tolerance of Measured Plane to Datum Plane
图 3-20 被测平面对基准平面的平行度公差

Parallelism tolerance of a line related to a datum plane, the tolerance zone is limited by two parallel planes with a distance of t and parallel to the datum plane. The direction of tolerance zone is fixed. It is qualified during the extracted median line shall be contained between two parallel planes 0.01mm apart, which are parallel to the datum plane A, as shown in Fig. 3-21.

被测直线对基准平面的平行度公差,公差带为间距等于公差值 t 且平行于基准平面的两平行平面所限定区域,公差带的方向是固定的。如图 3-21 所示,被测直线应限定在间距等于 0.01mm 且平行于基准平面的两平行平面之间。

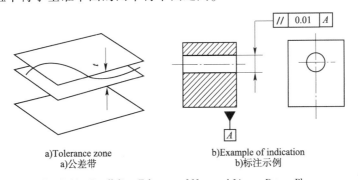

a)Tolerance zone a)公差带   b)Example of indication b)标注示例

Fig. 3-21 Parallelism Tolerance of Measured Line to Datum Plane
图 3-21 被测直线对基准平面的平行度公差

Parallelism tolerance of a line related to a datum line, the tolerance zone is limited by a cylinder of diameter t, parallel to the datum, if the tolerance value is preceded by the symbol ϕ. The direction of tolerance zone is fixed. It is qualified during the extracted median line shall be within a cylindrical zone of diameter 0.03mm, parallel to the datum shaft A, as shown in Fig. 3-22.

被测直线对基准直线的平行度，如果公差值前加了 ϕ，其公差带为直径等于公差值 t 且轴线平行于基准直线的圆柱面所限定区域，公差带的方向是固定的。如图 3-22 所示，被测孔的实际轴线应限定在直径等于 0.03mm 且平行于基准轴线的圆柱面内。

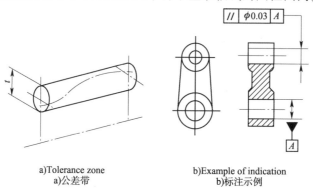

a)Tolerance zone
a)公差带

b)Example of indication
b)标注示例

Fig. 3-22 Parallelism Tolerance of Measured Line to Datum Line
图 3-22 被测直线对基准直线的平行度公差

3.4.2 Perpendicularity Tolerances 垂直度公差

Perpendicularity tolerance of a plane related to a datum plane, the tolerance zone is limited by two parallel planes a distance t apart and perpendicular to the datum plane. The direction of tolerance zone is fixed. It is qualified during the extracted surface shall be contained between two parallel planes 0.08mm apart, which are perpendicular to the datum plane A, as shown in Fig. 3-23.

被测平面对基准平面的垂直度，其公差带为间距等于公差值 t 且垂直于基准平面的两平行平面所限定的区域，公差带的方向是固定的。如图 3-23 所示，实际表面应限定在间距等于 0.08mm 且垂直于基准平面的两平行平面之间。

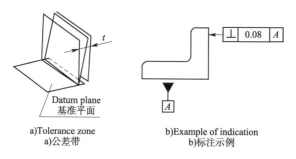

a)Tolerance zone
a)公差带

b)Example of indication
b)标注示例

Fig. 3-23 Perpendicularity Tolerance of Measured Plane to Datum Plane
图 3-23 被测平面对基准平面的垂直度公差

Perpendicularity tolerance of a line related to datum plane, the tolerance zone is limited by a cylinder of diameter t perpendicular to the datum, if the tolerance value is preceded by the symbol ϕ. The direction of tolerance zone is fixed. It is qualified during the extracted median line of the cylinder shall be within a cylindrical zone of diameter 0.01mm, which is perpendicular to the datum plane A, as shown in Fig. 3-24.

被测直线对基准平面的垂直度,如果公差值前加了 φ,其公差带为直径等于公差值 t 且轴线垂直于基准平面的圆柱面所限定区域,公差带的方向是固定的。如图 3-24 所示,被测孔的实际轴线应限定在直径等于 0.01mm 且垂直于基准平面的圆柱面内。

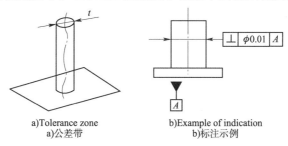

Fig. 3-24　Perpendicularity Tolerance of Measured Line to Datum Plane
图 3-24　被测直线对基准平面的垂直度公差

3.4.3　Angularity Tolerances 倾斜度公差

Angularity tolerance of a line related to datum line, line and datum line in the same plane: the tolerance zone is limited by two parallel planes a distance t apart and inclined at the specified angle to the datum. The direction of tolerance zone is fixed. It is qualified during the extracted median line shall be contained between two parallel planes 0.08mm apart that are inclined at a theoretically exact angle of 60° to the common datum straight line A-B, as shown in Fig. 3-25.

被测直线对基准直线的倾斜度,当被测直线与基准直线在同一平面上时,公差带为间距等于公差值 t 的两平行平面所限定的区域,该两平行平面按给定角度倾斜于基准轴线。公差带的方向是固定的。如图 3-25 所示,被测孔的实际轴线应限定在间距等于 0.08mm 的两平行平面之间,两平行平面按理论正确角度 60°倾斜于基准轴线 A-B。

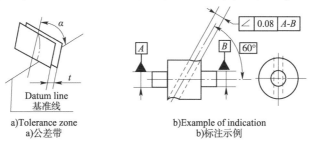

Fig. 3-25　Angularity Tolerance of Measured Line to Datum Line
图 3-25　被测直线对基准直线的倾斜度公差

3.5　Location Tolerances 位置公差

Location tolerance is the allowable variation of the location of the related feature to the datum feature. The location tolerance zone has a correct location relative to the datum, which is determined by the theoretical correct size, in which the theoretical correct size of coaxiality and symmetry is zero. The location tolerance has the function of controlling the location, direction and shape of the measured feature. Location tolerances are used to limit the location deviation of related feature respected to datum, which include coaxiality (concentricity), symmetry and position. The toleranced features of location tolerances include points, straight lines, and planes, and the datum tolerances mainly include straight lines and planes.

位置公差是关联要素对基准要素在位置上允许的变动量。位置公差带相对于基准具有正确的位置。位置度的公差带位置由理论正确尺寸确定,同轴度和对称度的理论正确尺寸为零。位置公差具有综合控制被测要素位置、方向和形状的功能。位置公差用于限制相对于基准的相对特征的位置偏差,包括同轴度(同心度)、对称度和位置度公差。位置公差的被测特征包括点、直线和平面,主要用于直线和平面。

Coaxiality (concentricity) and symmetry are special location tolerances, which employ the same tolerance concept but apply to different geometries. The tolerance feature of coaxiality tolerances is the shaft with a rotational feature such as cylinder, cone, etc, and the datum is also a shaft. And coaxial tolerance is used to control the error of coincidence degree between the toleranced shaft and the datum shaft. Concentricity is used to control the error of concentric degree between the toleranced center point and the datum center point. Symmetry tolerances control features constructed about a center point, a shaft or a median plane. The toleranced features which applied location tolerance should be keep the exact location relationships between toleranced feature and the datum feature(s) specified on engineering drawings, and the exact location relationship between toleranced feature and the datum feature(s) is determined by theoretical exact sizes (TED). The TED value is zero for coaxial (concentricity) and symmetry tolerances, and the TED need not indicated on engineering drawings.

同轴度(同心度)和对称度是特殊的位置公差,它们采用相同的公差概念,但适用于不同的几何形状。同轴度公差的被测要素特征是圆柱、圆锥等旋转特征的轴线,基准也是轴线。同轴度公差用于控制被测轴线与基准轴线重合度的误差。同心度用于控制被测中心点与基准中心点的同心度误差。对称度公差控制围绕中心点、轴或中间平面构造的特征误差。采用位置公差的被测要素应保持被测要素与工程图样上规定的基准要素之间的确切位置关系,被测要素与基准要素之间的准确位置关系由理论上的正确尺寸确定。同轴度(同心度)和对称度公差的理论正确尺寸值为0,在工程图纸上无须标明。

Location tolerances is related to datum(s), the orientation and position of tolerance zones are determined by datum. Location tolerances have the function to limit the deviation of location, orientation and form. Such as the position tolerance of a plane can limit the flatness error and orientation error related to datum of this plane; the coaxiality tolerance can limit the straightness error and parallelism error related to datum shaft of the toleranced shaft. So usually, under the premise of functional requirements, the orientation tolerances and form tolerances are not applied during the location tolerances are applied. If necessary, the form tolerances or orientation tolerances with high precision is needed, then the form tolerances or orientation tolerances can be applied, however, the tolerance value of form tolerance and orientation tolerance should be smaller than the location tolerance value.

位置公差是与基准相关联的,公差带的方向和位置依据基准来确定。位置公差具有限制位置、方向和形状误差的功能。如平面的位置公差可以限制与该平面对应的基准有关的平面度误差和方向误差;同轴度公差可以限制与被测轴线的基准轴线有关的直线度误差和平行度误差。因此,在满足功能要求的前提下,通常不采用方向公差和形状公差。如果需要高精度的形状公差或方向公差,则可采用形状公差或方向公差,但形状公差和方向公差的公差值应小于位置公差值。

On the premise of ensuring the functional requirements, after the position tolerance is given for the measured feature, the direction tolerance or shape tolerance will be given only when there are further requirements for the direction accuracy and shape accuracy. The direction tolerance value must be less than the position tolerance value, and the form tolerance value must be less than the orientation tolerance value. For example, as shown in Fig. 3-26, 0.05mm position tolerance, 0.03mm parallelism tolerance and 0.01mm flatness tolerance are given for the measured plane.

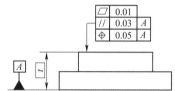

Fig. 3-26 Location Tolerance, Parallelism Tolerance and Flatness Tolerance are Given at the same time

图 3-26 同时给出位置公差、方向公差和形状公差

在保证功能要求的前提下,对被测要素给出位置公差后,仅在对其方向精度和形状精度有进一步要求时,才另行给出方向公差或形状公差,而方向公差值必须小于位置公差值,形状公差值必须小于方向公差值。如图 3-26 所示,对被测平面同时给出 0.05mm 位置度公差、0.03mm 平行度公差和 0.01mm 平面度公差。

3.5.1 Concentricity and Coaxiality Tolerances 同心度和同轴度公差

Concentricity tolerance of a point, the tolerance zone is limited by a circle of diameter t, the tolerance value shall be preceded by the symbol ϕ, the center of the circular tolerance zone coincides with the datum point. The location of tolerance zone is fixed. It is qualified during the extracted center of the inner circle in any cross-section shall be within a circle of diameter 0.01mm, concentric with datum point A defined in the same cross-section, as shown in Fig. 3-27.

点的同心度,公差带为直径等于公差值 t 的圆周所限定的区域,公差值前应加 ϕ,圆周的圆心与基准点重合。公差带的位置是固定的。如图 3-27 所示,在任一截面内,外圆的实际中心点应限定在直径等于 0.01mm 且以基准点 A 为圆心的圆周内。

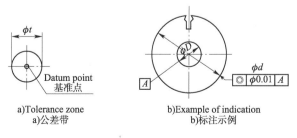

a) Tolerance zone
a) 公差带

b) Example of indication
b) 标注示例

Fig. 3-27 Concentricity Tolerance of a Point
图 3-27 同心度公差

Coaxiality tolerance of a shaft, the tolerance zone is limited by a cylinder of diameter t, the tolerance value is preceded by the symbol ϕ. The shaft of the cylindrical tolerance zone coincides with the datum. The location of tolerance zone is fixed. It is qualified during the extracted median line of the tolerance cylinder shall be within a cylindrical zone of diameter 0.08, the shaft of which is the common datum straight line A-B, as shown in Fig. 3-28.

线的同轴度,其公差带为直径等于公差值 t,公差值前应加 ϕ,且轴线与基准轴线重合的圆柱面所限定的区域。公差带的位置是固定的。如图 3-28 所示,被测圆柱面的实际轴线应限定在直径等于 0.08mm 且轴线与基准轴线 A-B 重合的圆柱面内。

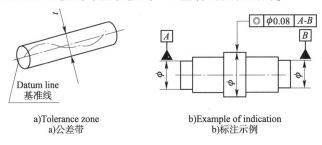

Fig. 3-28　Coaxiality Tolerance of a Shaft
图 3-28　同轴度公差

3.5.2　Symmetry Tolerances 对称度公差

Symmetry tolerance of a median plane, the tolerance zone is limited by two parallel planes a distance t apart, symmetrically disposed about the median plane, with respect to the datum. The location of tolerance zone is fixed. It is qualified during the extracted median surface shall be contained between two parallel planes 0.08mm apart, which are symmetrically disposed about datum plane A, as shown in Fig. 3-29.

被测平面对基准平面的对称度,公差带为间距等于公差值 t 且对称于基准中心平面的两平行平面所限定的区域。公差带位置是固定的。如图 3-29 所示,被测实际中心平面应限定在间距等于 0.08mm 且对称于基准中心平面 A 的两平行平面之间。

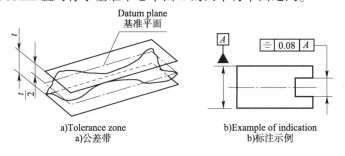

Fig. 3-29　Symmetry Tolerance of Measured Plane to Datum Plane
图 3-29　被测平面对基准平面的对称度公差

Symmetry tolerance of a median plane, the tolerance zone is limited by two parallel planes a distance t apart, symmetrically disposed about the median plane, with respect to the datum. The location of tolerance zone is fixed. It is qualified during the extracted median surface shall be contained between two parallel planes 0.08mm apart, which are symmetrically disposed about datum line A, as shown in Fig. 3-30.

被测平面对基准直线的对称度，公差带为间距等于公差值 t 且对称于基准轴线的两平行平面所限定的区域。公差带位置是固定的。如图 3-30 所示，被测实际中心平面应限定在间距等于 0.08mm 且对称于基准轴线 A 的两平行平面之间。

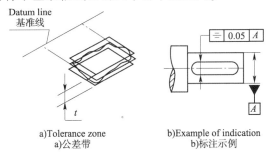

Fig. 3-30 Symmetry Tolerance of Measured Plane to Datum Line

图 3-30 被测平面对基准轴线的对称度公差

3.5.3 Position Tolerances 位置度公差

Position tolerance of a point, the tolerance zone is limited by a circle of diameter t, the tolerance value shall be preceded by the symbol ϕ, the center of the circular tolerance zone is fixed by theoretically exact sizes with respect to datum. The location of tolerance zone is fixed. It is qualified during the extracted center of the inner circle in any cross-section shall be within a circle of diameter 0.08mm, the center of which coincides with the theoretically exact position of the circle, with respect to datum plane A apart 68mm and datum plane B apart 100mm, as shown in Fig. 3-31.

点的位置度，公差带为直径等于公差值 t 的圆所限定的区域，公差值前应加 ϕ，该圆的中心的理论正确位置由基准和理论正确尺寸确定，公差带的位置是固定的。如图 3-31 所示，实际圆心应限定在直径等于 0.08mm 的圆内，圆的中心应处于由基准 A 和基准 B 以及理论正确尺寸 68mm 和 100mm 确定的位置。

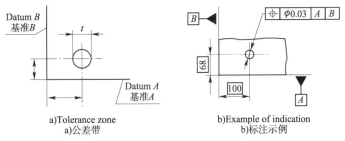

Fig. 3-31 Position Tolerance of a Point

图 3-31 点的位置度公差

Position tolerance of a line, the tolerance zone is limited by a cylinder of diameter t, the tolerance value is preceded by the symbol ϕ. The shaft of the tolerance cylinder is fixed by theoretically exact sizes with respect to datum. The location of tolerance zone is fixed. It is qualified during the extracted median line shall be within a cylindrical zone of diameter 0.08mm, the shaft of which coincides with the theoretically exact position of the considered hole, with respect to datum planes A, B and C, as shown in Fig. 3-32.

线的位置度，公差带为直径等于公差值 t 的圆柱面所限定的区域，公差值前应加 ϕ，该圆柱面的轴线的理论正确位置由基准和理论正确尺寸确定。公差带的位置是固定的。如图 3-32 所示,被测孔的实际轴线应限定在直径等于 0.08mm 的圆柱面内,圆柱面的轴线应处于由基准平面 A、B、C 和理论正确尺寸 68mm、100mm 确定的位置。

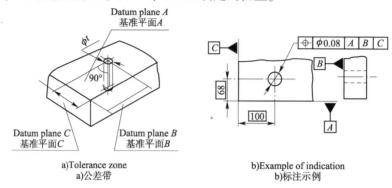

a)Tolerance zone
a)公差带

b)Example of indication
b)标注示例

Fig. 3-32　Position Tolerance of a Line
图 3-32　线的位置度公差

Position tolerance of a plane, the tolerance zone is limited by two parallel planes a distance t apart and symmetrically disposed about the theoretically exact position fixed by theoretically exact sizes with respect to the datum. The location of tolerance zone is fixed. It is qualified during the extracted surface is contained between two parallel planes 0.08mm apart that are symmetrically disposed about the theoretically exact position of the surface, with respect to datum plane A and datum shaft B, as shown in Fig. 3-33.

面的位置度,公差带为间距等于公差值 t 且对称于被测表面理论正确位置的两平行平面所限定的区域,理论正确位置由基准平面、基准轴线和理论正确尺寸以及理论正确角度确定。公差带位置是固定的。如图 3-33 所示,实际表面应限定在间距等于 0.08mm 且对称于被测表面理论正确位置的两平行平面之间,理论正确位置由基准平面 A、基准轴线 B 和理论正确尺寸 15mm 以及理论正确角度 105°确定。

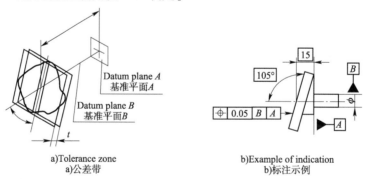

a)Tolerance zone
a)公差带

b)Example of indication
b)标注示例

Fig. 3-33　Position Tolerance of a Plane
图 3-33　面的位置度公差

3.6　Run-out Tolerances 跳动公差

Run-out tolerance is the maximum allowable run-out when the actual feature rotates around the datum shaft for one cycle or continuously. The run-out is the difference between the maximum

Chapter 3 Geometrical Tolerances 几何公差

and minimum indication values.

跳动公差是关联实际要素绕基准轴线回转一周或连续回转时所允许的最大跳动量。跳动量为指示表的最大与最小示值之差。

Run-out tolerances are defined by a special measuring method, which refers to the maximum difference of indication values when the actual surface rotated around a datum shaft. The toleranced features of run-out tolerance usually are cylindrical surface, cone surface and end plane. The datum feature is a shaft. Run-out includes circular run-out and total run-out. Circular run-out refers to the allowable variation of the toleranced feature rotation one circle in a cross-section relative to datum shaft. During measurement, the toleranced feature is rotated one round and the position of the indicator is fixed. According to the measuring direction, the circular run-out is divided into radial run-out, axial run-out and slant run-out. The measuring direction of radial run-out is perpendicular to the datum shaft, the measuring direction of axial run-out is parallel to the datum shaft, and the measuring direction of slant run-out is in any other angle except 0° and 90°. Total run-out refers to the allowable variation quantity of the whole toleranced feature related to datum shaft. During measurement, the toleranced feature rotates continually and the indicator moves straightly along the shaft direction. Total run-out is divided into radial total run-out and axial total run-out.

跳动公差由一种特殊的测量方法定义。该方法是指当实际表面绕基准轴旋转时,允许指示器读数的最大差异。跳动公差的被测要素为圆柱面、锥面或端面,基准要素为轴线。跳动包括圆跳动和全跳动。圆跳动是指被测要素在某个测量截面内相对于基准轴线旋转一周时所允许的变动量。测量时,被测要素旋转一圈,指示器的位置固定。根据测量方向,圆跳动分为径向圆跳动、轴向圆跳动和斜向圆跳动。径向跳动的测量方向与基准轴线垂直;轴向跳动的测量方向与基准轴线平行;斜跳动的测量方向为除0°和90°以外的任何角度。全跳动是指与基准轴线有关的整个被测要素的允许误差量。在测量过程中,被测要素不断旋转,指示器沿轴线方向直线移动。全跳动分为径向全跳动和轴向全跳动。

Run-out tolerances are related to a datum, and the orientation and position of the tolerance zone is fixed. Run-out tolerances are comprehensive tolerances, which can limit form, orientation and location error. Radial circular run-out tolerance can limit circularity error and concentricity error. Axial circular run-out tolerance can limit orientation error of a plane to shaft. Radial total run-out tolerance can limit cylindricity error and coaxiality error. Axial total run-out can limit perpendicular error of a plane to datum shaft and flatness error (axial total run-out is equivalent to perpendicularity of a plane to a datum shaft). As run-out is easily measured, the run-out tolerance is usually preferred specified under same functional condition during designing. For example, cylindricity tolerance should be specified to the cylinder surface fitted with rolling bearing, however, usually the run-out toleranced is used to replace the cylindricity tolerance by engineers in practical designing.

跳动公差与基准相关,公差带的方向和位置是固定的。跳动公差是一种综合公差,它可以限制形状误差、方向误差和位置误差。径向圆跳动公差可以限制圆度误差和同心度误差。轴向圆跳动公差可以限制平面对轴的定向误差。径向全跳动公差可以限制圆柱度误差和同轴度误差。轴向全跳动可限制平面对基准轴线的垂直误差和平面度误差(轴向全跳动相当于平面对基准轴线的垂直度)。由于跳动量容易测量,所以在设计时,通常在相同的功能条件下,优先规定跳动量公差。例如,对于装有滚动轴承的圆柱面,应规定圆柱度公差,但在实际设计中,工程师通常用被测的跳动公差代替圆柱度公差。

3.6.1　Circular Run-out Tolerances　圆跳动公差

Circular run-out tolerance-radial, the tolerance zone is limited within any cross-section perpendicular to the datum shaft by two concentric circles with a difference in radii of t, the centres of which coincide with the datum. The central location of tolerance zone is fixed. It is qualified during the extracted line in any cross-section plane perpendicular to datum shaft A shall be contained between two coplanar concentric circles with difference in radii of 0.1mm, as shown in Fig. 3-34.

径向圆跳动,公差带为在任一垂直于基准轴线的横截面内、半径差等于公差值 t、圆心在基准轴线上的两同心圆所限定的区域。公差带的中心位置固定。如图3-34所示,在任一垂直于基准轴线 A 的横截面内,被测圆柱面的实际圆应限定在半径差等于0.1mm且圆心在基准轴线 A 上的两个同心圆之间。

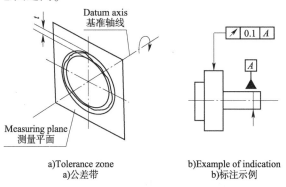

a)Tolerance zone　　　　b)Example of indication
a)公差带　　　　　　　　b)标注示例

Fig. 3-34　Circular Run-out Tolerance-Radial
图3-34　径向圆跳动公差

Circular run-out tolerance-axial, the tolerance zone is limited to any cylindrical section by circles a distance t apart lying in the cylindrical section, the shaft of which coincides with the datum. The central location of tolerance zone is fixed. It is qualified during line in any cylindrical section, the shaft of which coincides with datum shaft A, shall be contained between two circles with a distance of 0.1mm, as shown in Fig. 3-35.

轴向圆跳动,公差带为与基准轴线同轴的任一直径的圆柱截面上,间距等于公差值 t 的两个等径圆所限定的圆柱面区域。公差带的中心位置固定。如图3-35所示,在与基准轴线 A 同轴的任一直径的圆柱截面上,实际圆应限定在轴向距离等于0.1mm的两个等径圆之间。

Circular run-out tolerance-slant, the tolerance zone is limited within any conical section by two circles a distance t apart, the axes of which coincide with the datum. The width of the tolerance zone is normal to the specified geometry unless otherwise indicated. The central location of tolerance zone is fixed. It is qualified during the extracted line in any conical section, the shaft of which coincides with datum shaft A, shall be contained between two circles within the conical section with a distance of 0.1mm, as shown in Fig. 3-36.

斜向圆跳动，公差带为与基准轴线同轴的某一圆锥截面，间距等于公差值 t 的直径不相等的两个圆所限定的圆锥面区域，除非另有规定，测量方向应垂直于被测表面。公差带的中心位置固定。如图 3-36 所示，在与基准轴线 A 同轴的任一圆锥面，实际轮廓线应限定在素线方向间距等于 0.1mm 的直径不相等的两个圆之间。

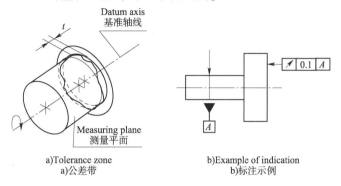

Fig. 3-35　Circular Run-out Tolerance-Axial

图 3-35　轴向圆跳动公差

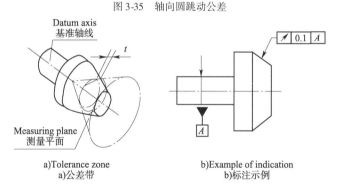

Fig. 3-36　Circular Run-out Tolerance-Slant

图 3-36　斜向圆跳动公差

3.6.2　Total Run-out Tolerances 全跳动公差

Radial total run-out tolerance, the tolerance zone is limited by two coaxial cylinders with a difference in radii of t, the axes of which coincide with the datum. The central location of tolerance zone is fixed. It is qualified during the extracted surface shall be contained between two coaxial cylinders with a difference in radii of 0.1mm and the axes coincident with the common datum straight line $A\text{-}B$, as shown in Fig. 3-37.

径向全跳动，公差带为半径差等于公差值 t 且轴线与基准轴线重合的两个圆柱面所限定的区域。公差带的中心位置固定。如图 3-37 所示，被测圆柱面的整个实际表面应限定在半径差等于 0.1mm 且轴线与公共基准轴线 $A\text{-}B$ 重合的两个圆柱面之间。

Axial total run-out tolerance, the tolerance zone is limited by two parallel planes a distance t apart and perpendicular to the datum. The central location of tolerance zone is fixed. It is qualified during the extracted surface shall be contained between two parallel planes 0.1mm apart, which are perpendicular to datum shaft A, as shown in Fig. 3-38.

轴向全跳动,公差带为半径差等于公差值 t 且垂直于基准轴线的两个平行平面所限定的区域。公差带的中心位置固定。如图 3-38 所示,被测实际表面应限定在间距等于 0.1mm 且垂直于基准轴线 A 的两个平行平面之间。

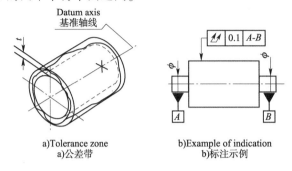

Fig. 3-37　Radial Total Run-out Tolerance

图 3-37　径向全跳动公差

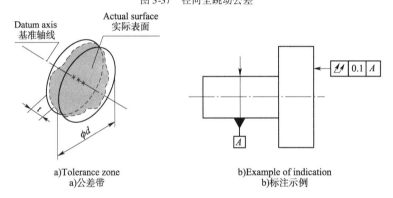

Fig. 3-38　Axial Total Run-out Tolerance

图 3-38　轴向全跳动公差

3.7　Tolerance Principles 公差原则

During designing mechanical components, both the size tolerances and geometrical tolerances are specified at same time for the key geometric features according to functional requirements. Moreover, the actual states of the geometric features of a workpiece is the result of comprehensive function between the size deviation and geometrical deviation, both the size deviation and geometrical deviation influence the fitting performance. So the relationship between size tolerance and geometrical tolerance should be explicit.

根据功能要求,在机械零件设计中对关键几何特征同时规定尺寸公差和几何公差,工件几何特征的实际状态是尺寸误差和几何误差综合作用的结果。因此,尺寸公差与几何公差的关系应明确。

Tolerance principle refers to the principle that defines relationship between size tolerances and geometrical tolerances. Tolerance principle is divided into independency principle and mutuality requirement. Mutuality requirement includes envelope requirement, maximum material requirement, minimum material requirement and reciprocity requirement. Due to the immaturity of reciprocity

Chapter 3 Geometrical Tolerances 几何公差

requirement, GB recommends careful use. In this text book, only independency principle, envelope requirement and maximum material requirement are introduced.

公差原则是指确定尺寸公差和几何公差之间关系的原则,公差原则分为独立原则和相关要求。相关要求包括包容要求、最大实体要求、最小实体要求和可逆要求。可逆要求由于不成熟,国标推荐谨慎使用。本教材只介绍了独立原则、包容要求和最大实体要求。

3.7.1 Terms and Definitions 术语和定义

3.7.1.1 External Function Size 体外作用尺寸

In the given length of the measured feature, the diameter or the width of the largest ideal surface connected with the actual inner surface or the smallest ideal surface connected with the actual external surface outside the body is called the external function size. For the related features, the shaft or central plane of the ideal plane must keep the geometric relationship given by the drawing with the datum. The external function size is formed by the actual size and geometric error of the measured features.

在被测要素的给定长度上,与实际内表面体外相接的最大理想面或与实际外表面体外相接的最小理想面的直径或宽度,称为体外作用尺寸。对于关联要素,该理想面的轴线或中心平面必须与基准保持图样给定的几何关系。体外作用尺寸是由被测要素的实际尺寸和几何误差综合形成的。

External function size of hole D_{fe} refers to, in the given length of toleranced feature, the width or diameter of the largest perfect feature that can be inscribed within the feature so that it just contacts the surface at highest points outside the workpiece. External function size of shaft d_{fe} refers to, in the given length of toleranced feature, the width or diameter of the smallest perfect feature that can be circumscribed about the feature so that it just contacts the surface at the highest points outside the workpiece.

在被测要素的给定长度上,与实际内表面体外相接的最大理想面的直径或宽度,称为孔的体外作用尺寸,用 D_{fe} 表示。在被测要素的给定长度上,实际外表面体外相接的最小理想面的直径或宽度,称为轴的体外作用尺寸,用 d_{fe} 表示作用。

Function size is the composite result of the deviations of local size and form, orientation, location of workpiece, and it is the actual fitting size when workpiece assembly. For related feature, the shaft or median plane should be kept the relationship with datum feature as indicated on engineering drawing.

作用尺寸是零件实际局部尺寸和形状、方向、位置误差的综合结果,是零件装配时的实际配合尺寸,其轴线或中心平面应保持与工程图样所示基准特性相一致的关系。

Because the external function size of a hole is smaller than the local size and the external function size of shaft is larger than the local size, the external function size affects the tightness degree after one shaft assembles with one hole. The external function size can be calculated by Equation (3-1) and Equation (3-2).

由于孔的体外作用尺寸小于局部尺寸,轴的体外作用尺寸大于局部尺寸,因此一孔一轴装配后,其体外作用尺寸影响着装配程度。孔和轴的体外作用尺寸可以通过式(3-1)和式(3-2)求得。

$$D_{fe} = D_a - f \tag{3-1}$$
$$d_{fe} = d_a + f \tag{3-2}$$

Where: f——the form, orientation or location deviation of derived feature.

式中:f——导出要素的形状、方向或位置误差。

3.7.1.2　Internal Function Size 体内作用尺寸

On the given length of the measured feature, the diameter or the width of the smallest ideal surface connected with the actual inner surface or the largest ideal surface connected with the actual outer surface body is called the internal function size. For the related features, the shaft or central plane of the ideal plane must keep the geometric relationship given by the drawing with the datum.

在被测要素的给定长度上,与实际内表面体内相接的最小理想面或与实际外表面体内相接的最大理想面的直径或宽度,称为体内作用尺寸。对于关联要素,该理想面的轴线或中心平面必须与基准保持图样给定的几何关系。

Internal function size of hole D_{fi} refers to, in the given length of toleranced feature, the width or diameter of the smallest perfect feature that can be circumscribed about the feature so that it just contacts the surface at highest points internal the workpiece. Internal function size of shaft d_{fi} refers to, in the given length of toleranced feature, the width or diameter of the largest perfect feature that can be inscribed within the feature so that it just contacts the surface at highest points internal the workpiece.

在被测要素的给定长度上,与实际内表面体内相接的最小理想面的直径或宽度,称为孔的体内作用尺寸,用 D_{fi} 表示。在被测要素的给定长度上,与实际外表面体内相接的最大理想面的直径或宽度,称为轴的体内作用尺寸,用 d_{fi} 表示。

The internal function size is also the composite result of the deviations of local size and form, orientation, location of workpiece. The internal function size of actual hole is larger than the actual size and the internal function size of shaft is smaller than the actual size, the internal function size can be calculated by Equation (3-3) and Equation (3-4).

体内作用尺寸也是由被测要素的实际尺寸和几何误差综合形成的。实际孔的体内作用尺寸大于该孔的实际尺寸,实际轴的体内作用尺寸小于该轴的实际尺寸。可以较直观地推导出轴的体内作用尺寸和孔的体内作用尺寸如式(3-3)和式(3-4)所示。

$$D_{fi} = D_a + f \tag{3-3}$$
$$d_{fi} = d_a - f \tag{3-4}$$

Where: f——the form, orientation or location deviation of derived feature.

式中:f——导出要素的形状、方向或位置误差。

3.7.1.3　Maximum Material Condition (MMC) and Maximum Material Size (MMS) 最大实体状态与最大实体尺寸

Maximum material condition (MMC) is the state of the considered feature of size in which the feature is everywhere at the limit size where the material of the feature is at its maximum, e. g. minimum hole diameter and maximum shaft diameter. Maximum material size (MMS) is the size defining the maximum material condition of a feature, i. e. the limit size where the material is at the maximum, e. g. maximum limit of size of a shaft or minimum limit size of a hole. The MMS of shaft is represented by d_M and MMS of hole is represented by D_M. The relationship of limit size and maximum material size is shown as Equation (3-5) and Equation (3-6).

最大实体状态(简称 MMC)是指实际要素在给定长度上处处位于尺寸极限之内并具有最大实体时的状态,或者说孔或轴具有允许的最多材料量时的状态。最大实体尺寸(简称 MMS)是指实际要素在最大实体状态下的极限尺寸。对于外表面,它为最大极限尺寸;对于内表面,它为最小极限尺寸。轴的最大实体尺寸用 d_M 表示,即轴的最大极限尺寸;孔的最大实体尺寸用 D_M,即孔的最小极限尺寸。两者分别如式(3-5)和式(3-6)所示。

$$d_M = d_{max} \quad (3-5)$$

$$D_M = D_{min} \quad (3-6)$$

3.7.1.4 Least Material Condition (LMC) and Least Material Size (LMS) 最小实体状态与最小实体尺寸

Least material condition (LMC) is the state of the considered feature of size in which the feature is everywhere at the limit size where the material of the feature is at its minimum, e. g. maximum hole diameter and minimum shaft diameter. Least material size (LMS) is the size defining the least material condition of a feature, i. e. the limit size where the material is at the minimum, e. g. minimum limit size of a shaft or maximum limit size of a hole. The LMS of shaft is represented by d_L, and LMS of hole is represented by D_L. The relationship of limit size and maximum material size is shown as Equation (3-7) and Equation (3-8).

最小实体状态(简称 LMC)是指实际要素在给定长度上处处位于尺寸极限之内并具有最小实体时的状态,或者说孔或轴具有允许的最少材料量时的状态。最小实体尺寸(简称 LMS)是指实际要素在最小实体状态下的极限尺寸。对于外表面,它为最小极限尺寸;对于内表面,它为最大极限尺寸。轴的最小实体尺寸用 d_L 表示,即轴的最小极限尺寸;孔的最小实体尺寸用 D_L,即孔的最大极限尺寸。两者分别如式(3-7)和式(3-8)所示。

$$d_L = d_{min} \quad (3-7)$$

$$D_L = D_{max} \quad (3-8)$$

3.7.1.5 Maximum Material Virtual Condition (MMVC) and Maximum Material Virtual Size (MMVS) 最大实体实效状态与最大实体实效尺寸

Maximum material virtual condition (MMVC) is a collective state of the considered feature of size in which the feature is at its MMC and its derived feature geometrical deviation is equal to its geometrical tolerance specified on engineering drawing. Maximum material virtual size (MMVS) is the external function size of the considered feature of size in which at MMVC. The MMVS of hole and shaft is represented by D_{MV} and d_{MV}, shown in Equation (3-9) and Equation (3-10).

在给定长度上,实际要素处于最大实体状态且其中心要素的几何误差等于给出公差值时的综合极限状态,称为最大实体实效状态(简称 MMVC)。最大实体实效状态下的体外作用尺寸,称为最大实体实效尺寸(简称 MMVS)。对于内表面,它为最大实体尺寸减几何公差值 t;对于外表面,它为最大实体尺寸加几何公差值 t。d_{MV} 为轴的最大实体实效尺寸代号,

D_{MV} 为孔的最大实体实效尺寸代号。根据定义,对于某一图样中的某一轴或孔的有关尺寸存在如式(3-9)和式(3-10)所示的关系。

$$D_{MV} = D_M - t \tag{3-9}$$

$$d_{MV} = d_M + t \tag{3-10}$$

Where: t——the geometrical tolerance of derived feature.

式中: t——相关几何特征的公差值。

3.7.1.6 Least Material Virtual Condition (LMVC) and Least Material Virtual Size (LMVS) 最小实体实效状态和最小实体实效尺寸

Least material virtual condition (LMVC) is a collective state of the considered feature of size in which the feature is at its LMC and its derived feature geometrical deviation is equal to its geometrical tolerance specified on engineering drawing. During a toleranced feature of size is related feature, MMVC and LMVC are limited by their orientation and location, i. e. the geometrical relationship should be kept with datum. Least material virtual size (LMVS) is the internal function size of the considered feature of size in which at LMVC. The LMVS of hole and shaft is represented by D_{LV} and d_{LV}, shown in Equation (3-11) and Equation (3-12).

在给定长度上,实际要素处于最小实体状态且其中心要素的几何误差等于给出公差值时的综合极限状态,称为最小实体实效状态(简称 LMVC)。最小实体实效状态下的体内作用尺寸,称为最小实体实效尺寸(简称 LMVS)。对于内表面,它为最小实体尺寸加几何公差值 t;对于外表面,它为最小实体尺寸减几何公差值 t。d_{LV} 为轴的最小实体实效尺寸代号,D_{LV} 为孔的最小实体实效尺寸代号。根据定义,对于某一图样中的某一轴或孔的有关尺寸存在如式(3-11)和式(3-12)所示的关系。

$$D_{LV} = D_L + t \tag{3-11}$$

$$d_{LV} = d_L - t \tag{3-12}$$

Where: t——the geometrical tolerance of derived feature.

式中: t——相关几何特征的公差值。

3.7.1.7 Boundary 边界

Boundary is the perfect limit envelope surface (cylinder or two parallel plane) specified at designing. If the toleranced feature is the individual feature, the boundary has no direction or position constraints. If the toleranced feature is the related feature, all the boundary should be kept the theory exact geometrical relationship related to datum. For the external surface, its boundary is equivalent to an internal surface with ideal shape; For the internal surface, its boundary is equivalent to an external surface with ideal shape. The width or diameter of the perfect limit envelope surface is the boundary size. The function of boundary is to control the deviation of size and geometrical (form, orientation or location). According to the functional and economy requirements, different boundaries (maximum material boundary, least material boundary, maximum material virtual boundary and least material boundary) are defined.

由设计给定的具有理想形状的极限包容面(极限圆柱或两平行平面)称为边界。单一要素的边界没有方向或位置的约束,而关联要素的边界则与基准保持图样上给定的几何关系。

对于外表面来说,它的边界相当于一个具有理想形状的内表面;对于内表面来说,它的边界相当于一个具有理想形状的外表面。该极限包容面的直径或宽度称为边界的尺寸。边界用于综合控制实际要素的尺寸和几何误差。根据零件的功能和经济性要求,可以给出最大实体边界、最小实体边界、最大实体实效边界和最小实体实效边界。

Maximum material boundary: the limit envelope surface of corresponding to maximum material condition.

最大实体边界:尺寸为最大实体尺寸的边界称为最大实体边界。

Least material boundary: the limit envelope surface of corresponding to least material condition.

最小实体边界:尺寸为最小实体尺寸的边界称为最小实体边界。

Maximum material virtual boundary: the limit envelope surface of corresponding to maximum material virtual condition.

最大实体实效边界:尺寸为最大实体实效尺寸的边界称为最大实体实效边界。

Least material virtual boundary: the limit envelope surface of corresponding to least material virtual condition.

最小实体实效边界:尺寸为最小实体实效尺寸的边界称为最小实体实效边界。

Example 3-1 As shown in Fig. 3-39a) and b), the diameter of the shaft and hole parts is measured as $\phi16$, and the linearity error of its shaft is 0.02mm. As shown in Fig. 3-39c) and d), the diameter of the shaft and hole parts is measured as $\phi16$, and the vertical error of the shaft is 0.2mm. The maximum material size, the least material size, the in external function size, the internal function size, the maximum material virtual size and the least material virtual size should be calculated.

例3-1 按图3-39a)、b)加工轴、孔零件,测得直径尺寸为$\phi16$,其轴线的直线度误差为0.02mm;按图3-39c)、d)加工轴、孔零件,测得直径尺寸为$\phi16$,其轴线的垂直度误差为0.2mm。试求出4种情况的最大实体尺寸、最小实体尺寸、体外作用尺寸、体内作用尺寸、最大实体实效尺寸和最小实体实效尺寸。

Solution:

解:

(1) Process parts according to Fig. 3-39a).

(1) 按图3-39a)加工零件。

$$d_M = d_{max} = 16$$
$$d_L = d_{min} = 16 + (-0.07) = 15.93$$
$$d_{fe} = d_a + f = 16 + 0.02 = 16.02$$
$$d_{fi} = d_a - f = 16 - 0.02 = 15.98$$
$$d_{MV} = d_M + t = 16 + 0.04 = 16.04$$
$$d_{LV} = d_L - t = 15.93 - 0.04 = 15.89$$

(2) Process parts according to Fig. 3-39b).

(2) 按图3-39b)加工零件。

$$D_M = D_{\min} = 16 + 0.05 = 16.05$$
$$D_L = D_{\max} = 16 + 0.12 = 16.12$$
$$D_{fe} = D_a - f = 16 - 0.02 = 15.98$$
$$D_{fi} = D_a + f = 16 + 0.02 = 16.02$$
$$D_{MV} = D_M - t = 16.05 - 0.04 = 16.01$$
$$D_{LV} = D_L + t = 16.12 + 0.04 = 16.16$$

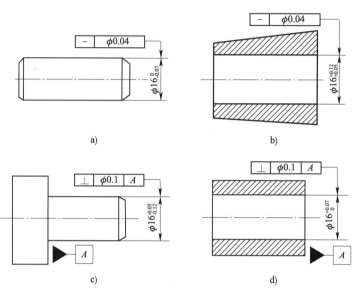

Fig. 3-39　Parts Drawing
图 3-39　零件图

(3) Process parts according to Fig. 3-39c).

(3) 按图 3-39c) 加工零件。

$$d_M = d_{\max} = 16 - 0.05 = 15.95$$
$$d_L = d_{\min} = 16 - 0.12 = 15.88$$
$$d_{fe} = d_a + f = 16 + 0.2 = 16.2$$
$$d_{fi} = d_a - f = 16 - 0.2 = 15.8$$
$$d_{MV} = d_M + t = 15.95 + 0.1 = 16.05$$
$$d_{LV} = d_L - t = 15.88 - 0.1 = 15.78$$

(4) Process parts according to Fig. 3-39d).

(4) 按图 3-39d) 加工零件。

$$D_M = D_{\min} = 16$$
$$D_L = D_{\max} = 16 + 0.07 = 16.07$$
$$D_{fe} = D_a - f = 16 - 0.2 = 15.8$$
$$D_{fi} = D_a + f = 16 + 0.2 = 16.2$$
$$D_{MV} = D_M - t = 16 - 0.1 = 15.9$$
$$D_{LV} = D_L + t = 16.07 + 0.1 = 16.17$$

Chapter 3　Geometrical Tolerances 几何公差

3.7.2　Independency Principle 独立原则

Independency principle refers to each specified dimensional or geometrical requirement on a drawing being met independently. It is the basic principle that the relationship of size tolerances and geometrical tolerances should conforme to. When independency principle is adopted, the size tolerances specified on drawing are only to control the size deviation and do not to control the geometrical deviations; and at same time, the geometrical tolerances are only to control the geometrical deviations and do not to control the size deviations. There are no additional symbols indicated on drawings when independency principle is adopted.

独立原则是指图纸上的每一个规定的尺寸或几何要求都满足相对独立的要求。独立原则是尺寸公差与几何公差关系应遵循的基本原则。当采用独立性原则时,图纸上规定的尺寸公差只控制尺寸误差,不控制几何误差;同时,几何公差只控制几何误差,不控制尺寸误差。当采用独立性原则时,图纸上无附加符号。

Fig. 3-40 is an example of independency principle used for individual feature. After machining, the local size of extracted feature should be kept between 30.979 ~ 30.000 mm, and the straightness deviation should not be more than 0.01mm, and the circularity deviation should not be more than 0.005mm. And the shaft is qualified during the two conditions are satisfied simultaneously.

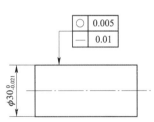

Fig. 3-40　Independency Principle is Used for Single Feature

图 3-40　独立原则应用于单一要素

图 3-40 是用于单个特征的独立性原则的示例。加工后,提取特征的局部尺寸应保持在 30.979 ~ 30.000mm,直线度误差不大于 0.01mm,圆柱度误差不大于 0.005mm。同时满足这两个条件时,轴是合格的。

For the toleranced feature which independency principle is used, the local size of extracted feature and the geometrical deviation should be measured respectively. Usually the local size is measure by two-point method and the geometrical deviation is measured by general or special measured tools. Independency principle is the essential tolerance principle, so it is widely used. Except there is obvious advantage to adopt mutuality principles, usually independency principle is used to specify the size tolerances and geometrical tolerances.

对于采用独立原则的被测特征,应分别测量提取特征的局部尺寸和几何误差。通常采用两点法测量局部尺寸,用通用或专用测量工具测量几何误差。独立原则是最基本的公差原则,因而被广泛应用。除了采用互换性原则具有明显的优点外,通常采用独立原则来规定尺寸公差和几何公差。

3.7.3　Envelope Requirement 包容要求

The envelope requirement, also called Taylor principle, specifies that the surface of an individual feature of size should not violate the imaginary envelope of perfect (geometrical ideal) form at maximum material size, i.e. the extracted integral feature should not exceed the maximum material boundary, and its extracted local size should not exceed the least material size. For a hole (internal feature of size), the external function size should be equal to or more than the maximum material size and the extracted local size should be equal to or less than the least material size. For a shaft (external feature of size), the external function size should be equal to or less than the maximum material size and the extracted local size should be equal to or more than the least material size. The envelope requirement can be indicated either by the symbol Ⓔ placed after the linear tolerance or by reference to an appropriate standard which invokes the envelope requirement.

包容要求适用于单一要素，例如圆柱面或两平行平面。仅对零件要素本身提出形状公差要求的要素，例如提出直线度要求的轴线。包容要求表示实际要素应遵守其最大实体边界，其局部实际尺寸不得超出最小实体尺寸。所谓遵守最大实体边界是指设计时应用边界尺寸即最大实体尺寸的边界，来控制被测要素的实际尺寸和形状误差的综合结果，要求该要素的实际轮廓不得超出这个边界（即体外作用尺寸不超出最大实体尺寸），并且实际尺寸不得超出最小实体尺寸。采用包容要求的单一要素应在其尺寸极限偏差或公差代号之后加注符号Ⓔ。

When an individual feature adopts the envelope requirement, the actual size and shape error of the feature depend on each other within the maximum solid boundary, and the allowable shape error value completely depends on the actual size. Therefore, if the actual size of the shaft or hole is the maximum physical size everywhere, the shape error must be zero to be qualified. For example, the drawing annotation in Fig. 3-41 indicates that the actual contour of an individual feature shaft shall not exceed the maximum solid boundary with a boundary size of 20mm, that is, the external function size of the shaft shall not be greater than the maximum material size of 20mm (the maximum limit size of the shaft). The actual size of the shaft shall not be less than the minimum material size of 19.987mm (the minimum limit size of the shaft). Because the shaft is limited by the maximum material boundary, when the shaft is in the maximum material state, there is no shape error; when the shaft is in the minimum material state, the allowable value of the shaft straightness error can reach $\phi 0.013$mm, and the cross-section shape of the shaft set in the figure is correct. However, in most cases, the shaft is in a certain actual size and has shape error (including shaft straightness). In this case, because the actual size deviates from the maximum material state, the shape tolerance value of the shaft can be compensated from this deviation, and the deviation can be compensated as much as possible. In other words, the shape tolerance value of the shaft is equal to the compensation value. At present, the actual size of the shaft is 19.998 mm, and the shape tolerance values of the shaft are listed in Tab. 3-2.

单一要素采用包容要求时，在最大实体边界范围内，该要素的实际尺寸和形状误差相互依赖，所允许的形状误差值完全取决于实际尺寸的大小。因此，若轴或孔的实际尺寸处处皆为最大实体尺寸，则其形状误差必须为0，才能合格。例如图3-41的图样标注，表示单一要素轴的实际轮廓不得超出边界尺寸为ϕ20mm的最大实体边界，即轴的体外作用尺寸应不大于20mm的最大实体尺寸(轴的最大极限尺寸)。轴的实际尺寸应不小于19.987mm的最小实体尺寸(轴的最小极限尺寸)。由于轴受到最大实体边界的限制，当轴处于最大实体状态时，不允许存在形状误差；当轴处于最小实体状态时，其轴线直线度误差允许值可达到ϕ0.013mm。假设图中轴的横截面形状正确。但是，在大部分情况下，轴是有某一实际尺寸的,而且具有形状误差(包括轴线直线度),这时,由于实际尺寸偏离了最大实体状态,轴的形状公差值可以从这个偏离中得到补偿。偏离多少就补偿多少,不偏离不补偿。也就是说,轴的形状公差值等于补偿值。现假设轴的实际尺寸为 19.998mm,将轴的形状公差值列于表3-2。

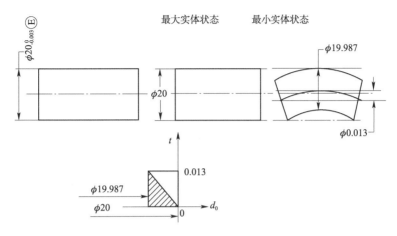

Fig. 3-41　Examples of Application of Independency Principle

图 3-41　包容要求应用举例

Tolerance Value of Shape Required by Inclusion　　　　　　　　　　Tab. 3-2

包容要求的形状公差值　　　　　　　　　　表 3-2

Type 类型	Maximum material condition 最大实体状态	Actual size 实际尺寸	Least material condition 最小实体状态	Boundary: Maximum material boundary 边界:最大实体边界
Size 尺寸	$\phi 20$	$\phi 19.998$	$\phi 19.987$	$d_M = \phi 20$
Tolerance value 形状公差值	0	Compensation value: 补偿值: $\phi 20 - \phi 19.998 = \phi 0.002$	$\phi 0.013$	—

When the actual feature deviates from the maximum material state, the tolerance of size is allowed to compensate the form tolerance, and the compensation amount depends on the deviation from the maximum material state. In Tab. 3-2, when the actual size is $\phi 19.998$mm, the compensation tolerance value is equal to $\phi 20$mm minus $\phi 19.998$mm, that is, $\phi 0.002$mm. The form tolerance referred to may be the straightness of the shaft, the straightness of the plain line, or the roundness, and the tolerance value is equal to the compensation value, which is 0.002mm.

当实际要素偏离最大实体状态时,包容要求允许尺寸公差补偿给形状公差,补偿量取决于偏离最大实体状态的多少。在表3-2中,当实际尺寸为 $\phi 19.998$mm,此时补偿形状公差值大小等于 $\phi 20$mm 减去 $\phi 19.998$mm 即 $\phi 0.002$mm,所指的形状公差可能是轴线直线度,也可能是素线直线度,还可能是圆度,公差值等于补偿值同为 0.002mm。

Judging the qualification conditions of parts from the meaning, the following parts qualification conditions can be obtained, such as Equation (3-13) and Equation (3-14) respectively for holes and shafts.

从含义上判断零件的合格条件,可以得出孔和轴的零件合格条件,分别如式(3-13)和式(3-14)所示。

$$\begin{cases} D_{fe} \geq D_M \\ D_a \leq D_L \end{cases} \quad 即 \quad \begin{cases} D_a - f \geq D_{min} \\ D_a \leq D_{max} \end{cases} \quad (3\text{-}13)$$

$$\begin{cases} d_{fe} \leq d_M \\ d_a \geq d_L \end{cases} \quad 即 \quad \begin{cases} d_a + f \leq d_{max} \\ d_a \geq d_{min} \end{cases} \quad (3\text{-}14)$$

Judging from the deviation from the maximum material state, the compensation value of shape tolerance can be calculated according to the degree of deviation from the maximum material state, as shown in Equation (3-15).

从偏离最大实体状态上判断,按偏离最大实体状态的程度可计算出形状公差的补偿值,如式(3-15)所示。

$$\begin{cases} d_{min} \leq d_a \leq d_{max} \quad 或 \quad D_{min} \leq D_a \leq D_{max} \\ f \leq t = 补偿值(\text{Compensation value}) \end{cases} \quad (3\text{-}15)$$

The compensation value in the formula is the deviation value from the maximum material state. For the shaft, the compensation value is equal to the maximum material size minus the actual size, i.e. $t = d_M - d_a$; for the hole, the compensation value is equal to the actual size of the hole minus the maximum physical size, that is, $t = D_a - D_M$.

式中的补偿值是偏离最大实体状态的偏离值。对于轴来说,补偿值等于最大实体尺寸减去实际尺寸,即 $t = d_M - d_a$;对于孔来说,补偿值等于孔的实际尺寸减去最大实体尺寸,即 $t = D_a - D_M$。

Taking the shaft as an example, the dynamic tolerance diagram of the shaft in Fig. 3-41 shows the rule that the straightness tolerance t of the shaft changes with the actual size d_a of the shaft. This tolerance value is the compensation value of the shaft relative to each actual size. As long as the error value falls in the shadow part of the figure, the shape tolerance of the shaft is qualified. When the parts are qualified, any of the above methods can be used, and the second method is relatively simple.

以轴为例,图3-41中下面的图为轴的动态公差图,该图表示轴线直线度公差值 t 随轴的实际尺寸 d_a 变化的规律,这个公差值就是轴相对于每个实际尺寸的补偿值。只要误差值落在图中的阴影部分,轴的形状公差就是合格的。判断零件合格时,上述任一种方法都可以,第二种方法比较简单。

The envelope requirement is often used to ensure the fitting properties of the hole and shaft, especially the precision fit requirement with small fitting tolerance. The maximum solid boundary is used to ensure the minimum clearance or maximum interference. For example, in the clearance positioning fit between the ϕ20H7 hole and the ϕ20h6 shaft, the minimum clearance required is zero, which is guaranteed by the maximum material boundary of the hole and the shaft respectively, without interference due to the shape error of the hole and shaft. When the envelope requirement is adopted, the upper deviation value of the center shaft of the base hole system is the minimum clearance or the maximum interference, and the lower deviation value of the middle hole of the base hole system is the minimum clearance or maximum interference. It should be pointed out that for the fit that the maximum interference is not strict and the minimum interference must be guaranteed, the envelope requirement is not necessary for the hole and shaft, because the minimum interference depends on the actual size of the hole and shaft, and is controlled by the least material size of the hole and shaft, rather than by their maximum material boundary. After the dimensional tolerance of a single feature is given according to the envelope requirements, if there is a higher requirement for the shape accuracy of the feature, the shape tolerance value can be further given, which must be less than the given dimensional tolerance value.

包容要求常用于保证孔、轴的配合性质,特别是配合公差较小的精密配合要求,用最大实体边界保证所需要的最小间隙或最大过盈。例如都采用包容要求的 ϕ20H7 孔与 ϕ20h6 轴的间隙定位配合中,所需要最小间隙为 0 的间隙配合性质是通过孔和轴各自遵守最大实体边界来保证的,不会因为孔和轴的形状误差而产生过盈。采用包容要求时,基孔制配合中轴的上偏差数值即为最小间隙或最大过盈;基轴制配合中孔的下偏差数值即为最小间隙或最大过盈。应当指出,对于最大过盈要求不严而最小过盈必须保证的配合,其孔和轴不必采用包容要求,因为最小过盈的大小取决于孔和轴的实际尺寸,是由孔和轴的最小实体尺寸控制的,而不是有它们的最大实体边界控制的。按包容要求给出单一要素的尺寸公差后,若对该要素的形状精度有更高的要求,还可以进一步给出形状公差值,这形状公差值必须小于给出的尺寸公差值。

Example 3-2 A shaft is processed according to the size $\phi 50_{-0.05}^{0}$, and the size in the pattern is processed according to the inclusion requirements. After processing, the actual size of the shaft is measured $d_a = \phi 49.97 \text{mm}$, and the shaft straightness error $f_- = \phi 0.02 \text{mm}$. Judge whether the part is qualified.

例 3-2 按尺寸 $\phi 50_{-0.05}^{0}$ 加工一个轴,图样上该尺寸按包容要求加工,加工后测得该轴的实际尺寸 $d_a = \phi 49.97 \text{mm}$,其轴线直线度误差 $f_- = \phi 0.02 \text{mm}$,判断该零件是否合格。

Solution:
解:
Judge from the meaning:
从含义上判断:

$$d_{max} = \phi 50 \text{mm}, d_{min} = \phi 49.95 \text{mm}$$
$$d_{fe} = d_a + f_- = \phi 49.97 + \phi 0.02 = \phi 49.99 < d_M = d_{max} = \phi 50$$
$$d_a = \phi 49.97 > d_L = d_{min} = \phi 49.95$$

So the part is qualified.
故零件合格。
Judge from the deviation:
从偏离状态判断:

$$d_{max} = \phi 50 \text{mm}, d_{min} = \phi 49.95 \text{mm}$$
$$d_{min} = \phi 49.95 < d_a = \phi 49.97 < d_{max} = \phi 50$$
$$f_- = \phi 0.02 < t = d_M - d_a = \phi 50 - \phi 49.97 = \phi 0.03$$

So the part is qualified.

故零件合格。

3.7.4 Maximum Material Requirements 最大实体要求

When the clearances of hole and shaft fit together, whether they can be assembled freely and ensure the functional requirements usually depends on the combined effect of local actual size and geometric error in vitro. For example, the bolt holes on two flanges are assembled with the bolts holding them together, when the local actual sizes of the bolt holes and bolts reach the maximum material size and their geometric errors also reach the given geometric tolerance value, their assembly clearance is the minimum value; when their local actual sizes deviate from the maximum material size and reaches the least material size and the geometric error is zero, their assembly clearance is the maximum. Accordingly, if the local actual sizes of bolt holes and bolts deviate from their maximum material sizes in the direction of the minimum material sizes, they are free to be assembled even if their geometric error exceeds the given geometric position tolerance (but not beyond a certain limit). This assembly depends on the concept of the relationship between the local actual size of the part and the geometric error, which is the theoretical basis for establishing the maximum material requirement.

孔与轴间隙配合时,它们能否自由装配和保证功能要求,通常取决于局部实际尺寸和几何误差的体外综合效应。例如,两个法兰盘上的螺栓孔与固紧它们的螺栓相装配,当螺栓孔和螺栓的局部实际尺寸都达到最大实体尺寸,且它们的几何误差也都达到给定几何公差值时,它们的装配间隙为最小值;当它们的局部实际尺寸偏离最大实体尺寸而达到最小实体尺寸和几何误差为零时,它们的装配间隙为最大值。据此,如果螺栓孔和螺栓的局部实际尺寸向最小实体尺寸方向偏离其最大实体尺寸,即使它们的几何误差超出给定几何公差值(但不超出某一限度),它们也能自由装配。这种装配取决于结合零件的局部实际尺寸及几何误差之间关系的概念,就是建立最大实体要求的理论依据。

Only interchangeable feature are required for assembly, usually the maximum material requirement is adopted. Therefore, the maximum material requirement is generally used in situations where the main guarantee of assemblability is lower for other functions. In this way, dimensional tolerance can be fully used to compensate geometric tolerance, which is beneficial to manufacturing and inspection. Maximum material requirements can only be used for shaft and shape, direction and position tolerances of central surfaces. If this principle can be applied correctly in design, it will bring beneficial economic effect to production. For example, maximum material requirements are widely applied to the positional tolerances of through-holes arranged on the circumference of a bolted or screwed-connected disk part in order to make full use of the through-hole dimensional tolerances given in the drawing and obtain the best technical and economic benefits.

只要求装配互换的要素,通常采用最大实体要求。因此,最大实体要求一般主要用于保证可装配性、对其他功能要求较低的场合。这样可以充分利用尺寸公差补偿几何公差,有利于制造和检验。最大实体要求只能用于轴线及中心面的形状、方向与位置公差。设计时如能正确地应用此原则,将给生产带来经济效益。例如,用螺栓或螺钉连接的圆盘零件上圆周布置的通孔的位置度公差广泛采用最大实体要求,以便充分利用图样上给出的通孔尺寸公差,获得最佳的技术和经济效益。

3.7.4.1 The Meaning of the Maximum Material Requirement and the Method of Marking on the Drawing 最大实体要求的含义和在图样上的标注方法

Maximum material requirements apply to derived feature (central feature). Adherence to the maximum material virtual boundary. When the actual size deviates from the maximum material size, the geometric error value is allowed to exceed the given tolerance value. The maximum material requirement applies to both the measured features and the datum feature. The symbol required for the maximum features is Ⓜ. When applied to the measured features, it shall be marked after the tolerance value of geometric tolerance frame; when applied to the datum feature, it shall be marked after the datum letter code of the geometric tolerance box.

最大实体要求适用于导出要素(中心要素)。遵守最大实体实效边界,即控制被测要素的实际轮廓处于其最大实体实效边界之内,当实际尺寸偏离最大实体尺寸时,允许几何误差值超出给定的公差值。最大实体要求既适用于被测要素也适用于基准要素。最大实体要求的符号为Ⓜ。应用于被测要素时,在几何公差框格的公差值后标注;应用于基准要素时,在几何公差框格的基准字母代号后标注。

3.7.4.2 The Maximum Material Requirement Applies to the Measured Feature 最大实体要求应用于被测要素

When the maximum material requirement is applied to the measured feature, the actual contour of the measured feature should not exceed the maximum material virtual boundary at any given length, that is, its external action size should not exceed the maximum material effect size, and its local actual size should not exceed the maximum material size and the least material size.

最大实体要求应用于被测要素时,被测要素的实际轮廓在给定的长度上处处不得超出最大实体实效边界,即其体外作用尺寸不应超出最大实体实效尺寸,且其局部实际尺寸不超出最大实体尺寸和最小实体尺寸。

Maximum material requirement is applied to the measured feature, the geometric tolerance values of measured features are given in the state of the features in the largest material. When measured features of the actual contour deviate from its biggest material state, namely the deviation from its maximum material size, actual size geometric error values can be beyond the maximum material condition given geometrical tolerance value, namely the geometric tolerance values can be increased.

最大实体要求应用于被测要素时,被测要素的几何公差值是在该要素处于最大实体状态时给出的。当被测要素的实际轮廓偏离其最大实体状态,即其实际尺寸偏离最大实体尺寸时,几何误差值可超出在最大实体状态下给出的几何公差值,即此时的几何公差值可以增大。

When the given geometric tolerance value is zero, it is zero geometric tolerance. At this time, the maximum material virtual boundary of the measured feature is equal to the maximum material virtual boundary, and the maximum material virtual size is equal to the maximum material size.

Maximum material requirements apply primarily to related features, but can also apply to individual features.

当给出的几何公差值为 0 时,则为零几何公差。此时,被测要素的最大实体实效边界等于最大实体边界,最大实体实效尺寸等于最大实体尺寸。最大实体要求主要应用于关联要素,也可用于单一要素。

The qualification conditions of the parts are described below.

零件的合格条件如下所述。

Judging the qualification conditions of parts from the meaning, the following parts qualification conditions can be obtained, such as Equation (3-16) and Equation (3-17) respectively for holes and shafts.

从含义上推断孔和轴的零件合格条件,分别如式(3-16)和式(3-17)。

$$\begin{cases} D_{fe} \geq D_{MV} \\ D_M \leq D_a \leq D_L \end{cases} \quad 即 \quad \begin{cases} D_a - f \geq D_{min} - t \\ D_{min} \leq D_a \leq D_{max} \end{cases} \tag{3-16}$$

$$\begin{cases} d_{fe} \leq d_{MV} \\ d_L \leq d_a \leq d_M \end{cases} \quad 即 \quad \begin{cases} d_a + f \leq d_{max} + t \\ d_{min} \leq d_a \leq d_{max} \end{cases} \tag{3-17}$$

Judging from the deviation from the maximum material condition, the compensation value of tolerance can be calculated according to the degree of deviation from the maximum material condition, as shown in Equation (3-18).

从偏离最大实体状态上判断,按偏离最大实体状态的程度可计算出公差的补偿值,如式(3-18)所示。

$$\begin{cases} d_{min} \leq d_a \leq d_{max} \quad 或 \quad D_{min} \leq D_a \leq D_{max} \\ f \leq t = 给定值 + 补偿值(\text{Given value + Compensation value}) \end{cases} \tag{3-18}$$

The given value is the tolerance value given in the tolerance frame. The compensation value is the deviation value from the maximum material condition. For the shaft, the compensation value is equal to the maximum material size minus the actual size, i. e. $t = d_M - d_a$; for the hole, the compensation value is equal to the actual size of the hole minus the maximum material size, that is, $t = D_a - D_M$.

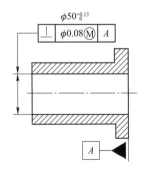

Fig. 3-42 Example of Applying the Maximum Material Requirement to the Measured Feature

图 3-42 最大实体要求应用于被测要素举例

式中的给定值是几何公差框格中给定的公差值,补偿值是偏离最大实体状态的偏离值。对于轴来说,补偿值等于最大实体尺寸减去实际尺寸,即 $t = d_M - d_a$;对于孔来说,补偿值等于孔的实际尺寸减去最大实体尺寸,即 $t = D_a - D_M$。

Maximum material requirements are applied to the measured features, as shown in Fig. 3-42. Set the actual size as $\phi 50.12$, and the shaft perpendicularity tolerance allowed is shown in Tab. 3-3.

最大实体要求应用于被测要素,如图3-42所示。设实际尺寸为 $\phi 50.12$,允许的轴线垂直度公差如表3-3所示。

Tolerance Value of the Maximum Material Requirement 最大实体要求的几何公差值

Tab. 3-3 表 3-3

Type 类型	Maximum material condition 最大实体状态	Actual size 实际尺寸	Minimum material condition 最小实体状态	Boundary: Maximum material virtual boundary 边界: 最大实体实效边界
Size 尺寸	$\phi 50$	$\phi 50.12$	$\phi 50.13$	$d_{MV} = \phi 49.92$
Tolerance value 公差值	$\phi 0.08$	Given value + Compensation value: 给定值 + 补偿值: $\phi 0.08 + (\phi 50.12 - \phi 50) = \phi 0.2$	Given value + Compensation value: 给定值 + 补偿值: $\phi 0.08 + (\phi 50.13 - \phi 50) = \phi 0.21$	—

Example 3-3 A hole is processed as shown in Fig. 3-42. The actual size of the hole measured after processing is $\phi 50.02$, and the error value of perpendicularity of the shaft is $\phi 0.09$. Determined whether the part is qualified.

例 3-3 按图 3-42 加工一个孔,加工后测得孔的实际尺寸为 $\phi 50.02$,轴的垂直度误差值 f 为 $\phi 0.09$,判断该零件是否合格。

Solution:

解:

Judge from the deviation:

从偏离状态判断:

$$D_{max} = \phi 50.13 \text{mm}, D_{min} = \phi 50 \text{mm}$$
$$D_{min} = \phi 50 < D_a = \phi 50.02 < D_{max} = \phi 50.13$$
$$f = \phi 0.09 < t = 给定值 + 补偿值 (\text{Given value + Compensation value})$$
$$= \phi 0.08 + (\phi 50.02 - \phi 50) = \phi 0.10$$

So the parts are qualified.

故零件合格。

3.7.5 Minimum Material Requirements 最小实体要求

The guarantee of the critical distance between adjacent features on the same part (such as the minimum wall thickness or the maximum distance) usually depends on the local actual size of the feature and the internal synthesis effect of the geometric error. For example, the wall thickness between two adjacent holes on a part is the minimum when the local actual size of both holes reaches the least material size and the position error between them also reaches the given position tolerance. Accordingly, if the local actual size of two holes deviates from its least material size towards the maximum material size, they can guarantee the minimum wall thickness even if their position error exceeds the given position degree tolerance value (but not beyond a certain limit). The concept that the critical distance depends on the relation between the local actual size and the shape and position error of the adjacent feature on the same part is the theoretical basis for establishing the minimum material requirement.

同一零件上相邻要素之间的临界距离(如最小壁厚或最大距离)的保证,通常取决于要素的局部实际尺寸和几何误差的体内综合效应。例如,对于零件上相邻两孔之间的壁厚,当两孔的局部实际尺寸都达到最小实体尺寸,且它们之间的位置误差也达到给定的位置度公差时,它们之间的壁厚为最小值。据此,如果两孔的局部实际尺寸向最大实体尺寸方向偏离其最小实体尺寸,即使它们的位置误差超出给定位置度公差值(但不超出某一限度),它们也能保证最小壁厚。这种临界距离取决于同一零件上相邻要素的局部实际尺寸及形位误差之间关系的概念,就是建立最小实体要求的理论依据。

Minimum material requirements are widely used to ensure the minimum wall thickness and control the maximum distance from the surface to the central features under the premise of obtaining the best technical and economic benefits. It is mainly used to limit the position variation of features, mostly used for position tolerance, and is suitable for derived (central) features.

在获得最佳技术和经济效益的前提下,最小实体要求广泛应用于保证最小壁厚和控制表面至中心要素的最大距离等功能要求。它主要用于限制要素的位置变动,多用于位置公差,适用于导出(中心)要素。

3.7.5.1 Definition of Minimum Material Requirements and Method of Marking on Drawings 最小实体要求的含义和在图样上的标注方法

Minimum material requirements apply to derived (central) feature. Following the minimum material virtual boundary, the actual contour of the measured feature is within its minimum material virtual boundary. When the actual size deviates from the least material size, the geometric error value is allowed to exceed the given tolerance value. The minimum material requirement applies to both the measured features and the datum features. The symbol required for the minimum material is Ⓛ. When applied to the measured features, the tolerance value of the geometric tolerance box of the features to be measured shall be marked. When applied to the datum feature, it shall be marked after the datum letter code of the geometric tolerance box.

最小实体要求适用于导出要素(中心要素)。被测要素遵守最小实体实效边界,其实际轮廓处于其最小实体实效边界之内。当实际尺寸偏离最小实体尺寸时,允许几何误差值超出给定的公差值。最小实体要求既适用于被测要素,也适用于基准要素。最小实体要求的符号为Ⓛ。应用于被测要素时,在被测要素几何公差框格的公差值后标注;应用于基准要素时,在几何公差框格的基准字母代号后标注。

3.7.5.2 Minimum Material Requirements Apply to the Measured Feature 最小实体要求应用于被测要素

When the minimum material requirement is applied to the measured feature, the actual contour of the measured feature should not exceed the minimum material virtual boundary at any given length, that is, the body action size should not exceed the least material virtual size, and the local actual size should not exceed the maximum material size and the least material size.

最小实体要求应用于被测要素时,被测要素的实际轮廓在给定的长度上处处不得超出最小实体实效边界,即其体内作用尺寸不应超出最小实体实效尺寸,且其局部实际尺寸不得超出最大实体尺寸和最小实体尺寸。

Minimum material requirements applied to the features being measured, the measured features of geometric tolerance values are given in minimum material virtual condition, when the actual contour measured features from its minimum material virtual condition, namely its actual size deviation minimum material sizes, geometric error values can be beyond the geometric tolerance is given under the minimum material virtual condition, namely the geometric tolerance values can be increased.

最小实体要求应用于被测要素时,被测要素的几何公差值是在该要素处于最小实体状态时给出的,当被测要素的实际轮廓偏离其最小实体状态,即其实际尺寸偏离最小实体尺寸时,几何误差值可超出在最小实体状态下给出的几何公差值,此时的几何公差值可以增大。

When the given geometric tolerance value is zero, it is zero geometric tolerance. The minimum material virtual boundary of the feature under test is equal to the minimum material boundary, and the minimum material virtual size is equal to the least material size. Minimum material requirements are applied primarily to related features and may also be applied to individual features.

当给出的几何公差值为0时,则为零几何公差。被测要素的最小实体实效边界等于最小实体边界,最小实体实效尺寸等于最小实体尺寸。最小实体要求主要应用于关联要素,也可用于单一要素。

The qualification conditions of the parts are described below.

零件的合格条件如下所述。

Judging the qualification conditions of parts from the meaning, the following parts qualification conditions can be obtained, such as Equation (3-19) and Equation (3-20) respectively for holes and shafts.

从含义上推断孔和轴的零件合格条件,分别如式(3-19)和式(3-20)。

$$\begin{cases} D_{\text{fi}} \leqslant D_{\text{LV}} \\ D_{\text{M}} \leqslant D_{\text{a}} \leqslant D_{\text{L}} \end{cases} \quad 即 \quad \begin{cases} D_{\text{a}} + f \leqslant D_{\max} + t \\ D_{\min} \leqslant D_{\text{a}} \leqslant D_{\max} \end{cases} \tag{3-19}$$

$$\begin{cases} d_{\text{fi}} \geqslant d_{\text{LV}} \\ d_{\text{L}} \leqslant d_{\text{a}} \leqslant d_{\text{M}} \end{cases} \quad 即 \quad \begin{cases} d_{\text{a}} - f \geqslant d_{\min} - t \\ d_{\min} \leqslant d_{\text{a}} \leqslant d_{\max} \end{cases} \tag{3-20}$$

Judging from the deviation from the least material condition, the compensation value of tolerance can be calculated according to the deviation from the maximum material condition, as shown in Equation (3-21).

从偏离最小实体状态上判断零件的合格条件,按偏离最小实体状态的程度可计算出公差的补偿值,如式(3-21)所示。

$$\begin{cases} d_{\min} \leqslant d_{\text{a}} \leqslant d_{\max} \quad 或 \quad D_{\min} \leqslant D_{\text{a}} \leqslant D_{\max} \\ f \leqslant t = 给定值 + 补偿值(\text{Given value} + \text{Compensation value}) \end{cases} \tag{3-21}$$

The given value is the tolerance value given in the tolerance frame. The compensation value is the deviation value from the least material condition. For the shaft, the compensation value is equal to the actual size minus the least material size, i.e. $t = d_{\text{a}} - d_{\text{L}}$; for the hole, the compensation value is equal to the least material size of the hole minus the actual size, that is, $t = D_L - D_a$.

式中的给定值是公差框格中给定的公差值,补偿值是偏离最小实体状态的偏离值。对于轴来说,补偿值等于实际尺寸减去最小实体尺寸,即 $t = d_a - d_L$;对于孔来说,补偿值等于孔的最小实体尺寸减去实际尺寸,即 $t = D_M - D_a$。

3.8 Selection of Geometrical Tolerance
几何公差的选用

3.8.1 Selection of Geometrical Characteristics 几何公差特征的选择

Following aspects should be considered when geometrical characteristics are selected. Structure properties of workpieces: to analyse the geometrical errors that maybe exist in the workpieces maching. For example, cylinder surface workpieces have cylindrical error; stepped shafts and holes have coaxial errors. Functional requirements of workpieces: to analysis the geometrical characteristics which affect the functional requirement. For example, the main errors which affect lathe spindle accuracy is the cylindrical error and coaxial error of the shaft neck in the front and the rear. The movement accuracy of lathe guide rail mainly is affected by straightness error of the guide rail. Properties of every geometrical characteristics: there are individual characteristics (straightness, flatness, circularity) and composite characteristics (cylindricity, location) in the 14 geometrical characteristics. It is preferred to select the composited characteristics to reduce the amount of characteristics on the engineer drawings. Measurement conditions: the measurement conditions include with or without measurement instrument, the measurement difficulty, measurement efficiency etc. In the case of satisfying the function requirements, the characteristics easily to be measured should be selected firstly.

在选择几何特征时,应考虑以下几个方面。工件的结构特性:分析被加工工件可能存在的几何误差。例如,圆柱面工件存在圆柱误差,配合轴和孔存在的同轴度误差。工件的功能要求:分析影响功能要求的几何特征。例如,影响车床主轴精度的主要误差是前后轴颈的圆柱度误差和同轴度误差。机床导轨的运动精度主要受导轨直线度误差的影响。每一个几何特征的性质:在14个几何特征中有各自的特征(直线度、平面度、圆度)和综合特征(圆柱度、位置)。最好选择复合特征,以减少工程图纸上的特征数量。测量条件因素:测量条件包括有无测量仪器、测量难度、测量效率等。在满足功能要求的情况下,应首先选择易于测量的特性。

The selection of datum mainly according to the function and design requirements of workpiece, and the aspects should be considered at the same time such as the principle of unifying datum and the structure properties of workpiece. Principle of unifying datum: the datum features should be the same with features of designing datum, location datum and the assembly datum. Among the three datum planes, the plane which has the largest effect on function or the location is most stable to the toleranced feature should be selected as primary datum.

基准的选择主要依据工件的功能和设计要求,同时考虑基准统一原则和工件的结构特性等因素。基准统一原则：基准特征应与设计基准、定位基准和装配基准相同。在3个基准面中，对功能影响最大或位置对公差特性最稳定的平面应作为基准面。

The envelope requirement is selected when the fit properties is very important, and maximum material requirement is selected when to ensure assemble freely, and minimum material requirement is selected when to ensure the minimum wall thickness or the maximum distance. Otherwise, the independency principle is selected.

当配合性能非常重要时，选择包容要求。在保证自由装配时选择最大实体要求。在保证最小壁厚和最大距离时选择最小实体要求，其余情况选择独立性原则。

3.8.2 Selection of Geometrical Tolerance Value 几何公差值的选择

Geometrical tolerance grades and tolerance values, in the Appendix B of *Geometrical tolerancing Geometrical tolerance for features without individual tolerance indications* (GB/T 1184—1996), the tolerance grades and values are specified for 11 geometrical tolerance characteristics, and the values series of position tolerance are specified, however, the tolerance grades are not specified. The two geometrical tolerance characteristics, i.e. profile of any line and profile of any surface, are not specified tolerance grades and values. There are 12 grades for the 11 geometrical characteristics which the tolerance grades are specified, shown in Tab. 3-4 to Tab. 3-7.

几何公差等级和公差值，《形状和位置公差　未注公差值》(GB/T 1184—1996)附录B规定了11个几何公差特征的公差等级和公差值，规定了位置公差的公差值，但未规定公差等级。两个几何公差特征，即线轮廓度和表面轮廓度，均未规定公差等级和公差值。规定了公差等级的11个几何特征共有12个等级，见表3-4至表3-7。

Geometrical Tolerance on Straightness and Flatness　　Tab. 3-4
直线度和平面度的公差值　　表3-4

Main parameter 主参数 L(mm)	Precision grades 精度等级											
	1	2	3	4	5	6	7	8	9	10	11	12
	Tolerance values 公差值(μm)											
≤10	0.2	0.4	0.8	1.2	2	3	5	8	12	20	30	60
>10~16	0.25	0.5	1.0	1.5	2.5	4	6	10	15	25	40	80
>16~25	0.3	0.6	1.2	2	3	5	8	12	20	30	50	100
>25~40	0.4	0.8	1.5	2.5	4	6	10	15	25	40	60	120
>40~63	0.5	1.0	2	3	5	8	12	20	30	50	80	150
>63~100	0.6	1.2	2.5	4	6	10	15	25	40	60	100	200
>100~160	0.8	1.5	3	5	8	12	20	30	50	80	120	250
>160~250	1.0	2	4	6	10	15	25	40	60	100	150	300
>250~400	1.2	2.5	5	8	12	20	30	50	80	120	200	400
>400~630	1.5	3	6	10	15	25	40	60	100	150	250	500
>630~1000	2	4	8	12	20	30	50	80	120	200	300	600

Continued 续上表

Main parameter 主参数 L(mm)	Precision grades 精度等级											
	1	2	3	4	5	6	7	8	9	10	11	12
	Tolerance values 公差值(μm)											
>1000~1600	2.5	5	10	15	25	40	60	100	150	250	400	800
>1600~2500	3	6	12	20	30	50	80	120	200	300	500	1000
>2500~4000	4	8	15	25	40	60	100	150	250	400	600	1200
>4000~6300	5	10	20	30	50	80	120	200	300	500	800	1500
>6300~10000	6	12	25	40	60	100	150	250	400	600	1000	2000

Geometrical Tolerance on Circularity and Cylindricity   Tab. 3-5
圆度和圆柱度的公差值   表3-5

Main parameter 主参数 $d(D)$(mm)	Precision grades 精度等级											
	1	2	3	4	5	6	7	8	9	10	11	12
	Tolerance values 公差值(μm)											
≤3	0.1	0.2	0.3	0.8	1.2	2	3	4	6	10	14	25
>3~6	0.1	0.2	0.4	1.0	1.5	2.5	4	5	8	12	18	30
>6~10	0.12	0.25	0.4	1.0	1.5	2.5	4	6	9	15	22	36
>10~18	0.15	0.25	0.5	1.2	2	3	5	8	11	18	27	43
>18~30	0.2	0.3	0.6	1.5	2.5	4	6	9	13	21	33	52
>30~50	0.25	0.4	0.6	1.5	2.5	4	7	11	16	25	39	62
>50~80	0.3	0.5	0.8	2	3	5	8	13	19	30	46	74
>80~120	0.4	0.6	1.0	2.3	4	6	10	15	22	35	54	87
>120~180	0.6	1.0	1.2	3.5	5	8	12	18	25	40	63	100
>180~250	0.8	1.2	2	4.5	7	10	14	20	29	46	72	115
>250~315	1.0	1.6	2.5	6	8	12	16	23	32	52	81	130
>315~400	1.2	2	3	7	9	13	18	25	36	57	89	140
>400~500	1.5	2.5	4	8	10	15	20	27	40	63	97	155

Geometrical Tolerance on Orientation   Tab. 3-6
方向公差的公差值   表3-6

Main parameter 主参数 L(mm)	Precision grades 精度等级											
	1	2	3	4	5	6	7	8	9	10	11	12
	Tolerance values 公差值(μm)											
≤10	0.4	0.8	1.5	3	5	8	12	20	30	50	80	120
>10~16	0.5	1.0	2	4	6	10	15	25	40	60	100	150

Chapter 3 Geometrical Tolerances 几何公差

Continued 续上表

Main parameter 主参数 L(mm)	Precision grades 精度等级											
	1	2	3	4	5	6	7	8	9	10	11	12
	Tolerance values 公差值(μm)											
>16~25	0.6	1.2	2.5	5	8	12	20	30	50	80	120	200
>25~40	0.8	1.5	3	6	10	15	25	40	60	100	150	250
>40~63	1.0	2	4	8	12	20	30	50	80	120	200	300
>63~100	1.2	2.5	5	10	15	25	40	60	100	150	250	400
>100~160	1.5	3	6	12	20	30	50	80	120	200	300	500
>160~250	2	4	8	15	25	40	60	100	150	250	400	600
>250~400	2.5	5	10	20	30	50	80	120	200	300	500	800
>400~630	3	6	12	25	40	60	100	150	250	400	600	1000
>630~1000	4	8	15	30	50	80	120	200	300	500	800	1200
>1000~1600	5	10	20	40	60	100	150	250	400	600	1000	1500
>1600~2500	6	12	25	50	80	120	200	300	500	800	1200	2000
>2500~4000	8	15	30	60	100	150	250	400	600	1000	1500	2500
>4000~6300	10	20	40	80	120	200	300	500	800	1200	2000	3000
>6300~10000	12	25	50	100	150	250	400	600	1000	1500	2500	4000

Geometrical Tolerance on Coaxial, Symmetry and Run-out   Tab. 3-7
同轴度、对称度和跳动的公差值   表3-7

Main parameter 主参数 L(mm)	Precision grades 精度等级											
	1	2	3	4	5	6	7	8	9	10	11	12
	Tolerance values 公差值(μm)											
≤1	0.4	0.6	1.0	1.5	2.5	4	6	10	15	25	40	60
>1~3	0.4	0.6	1.0	1.5	2.5	4	6	10	20	40	60	120
>3~6	0.5	0.8	1.2	2	3	5	8	12	25	50	80	150
>6~10	0.6	1.0	1.5	2.5	4	6	10	15	30	60	100	200
>10~18	0.8	1.2	2	3	5	8	12	20	40	80	120	250
>18~30	1.0	1.5	2.5	4	6	10	15	25	50	100	150	300
>30~50	1.2	2	3	5	8	12	20	30	60	120	200	400
>50~120	1.5	2.5	4	6	10	15	25	40	80	150	250	500
>120~250	2	3	5	8	12	20	30	50	100	200	300	600
>250~500	2.5	4	6	10	15	25	40	60	120	250	400	800

Continued 续上表

Main parameter 主参数 L(mm)	Precision grades 精度等级											
	1	2	3	4	5	6	7	8	9	10	11	12
	Tolerance values 公差值(μm)											
>500~800	3	5	8	12	20	30	50	80	150	300	500	1000
>800~1250	4	6	10	15	25	40	60	100	200	400	600	1200
>1250~2000	5	8	12	20	30	50	80	120	250	500	800	1500
>2000~3150	6	10	15	25	40	60	100	150	300	600	1000	2000
>3150~5000	8	12	20	30	50	80	120	200	400	800	1200	2500
>5000~8000	10	15	25	40	60	100	150	250	500	1000	1500	3000
>8000~10000	12	20	30	50	80	120	200	300	600	1200	2000	4000

For the economical reason, the geometrical tolerance value as large as possible is selected in the case satisfy the functional requirements. So far, there are no reliable methods to calculate geometrical tolerance values, so usually analogy method is used to select geometrical tolerance values. The following aspects should be considered when the geometrical tolerance values are selected.

几何公差值的选取,要从经济角度考虑,在满足功能要求的情况下,尽可能选取较大的公差值。到目前为止,还没有可靠的方法来计算几何公差值,所以一般都用类比法来选择几何公差值。在选择几何公差值时,应考虑以下几个方面。

The relationship of size tolerance and geometrical tolerance: usually, the values of form tolerance T_f, orientation tolerance T_0, location tolerance T_1 and size tolerance T_s should satisfy the Equation (3-22).

尺寸公差与几何公差的关系:形状公差 T_f、方向公差 T_0、位置公差 T_1 和尺寸公差 T_s 的值通常应满足式(3-22)所示的关系。

$$T_f < T_0 < T_1 < T_s \tag{3-22}$$

For example, if two parallel planes require geometrical tolerances, the flatness tolerance value should be smaller than the parallelism tolerance value, and the parallelism tolerance value should be smaller than the distance tolerance value between them.

例如,两个平行平面的几何公差,平面度公差值应小于平行度公差值,平行度公差值应小于它们之间的距离公差值。

The relationship of size tolerance and geometrical tolerance when envelope requirement is specified: the envelope requirement should be specified for the features of which the fit requirement should be satisfied strickly. During manufacturing, the form tolerance values T_f come from size tolerance T_s, the relationship is shown in equation (3-23). Usually take the $K = 25\% \sim 65\%$ for the size tolerance grade between IT5 ~ IT8.

规定包容要求时尺寸公差与几何公差的关系:对必须严格满足配合要求的特征,应规定包容要求。在制造过程中,形状公差值 T_f 的选取依据尺寸公差 T_s,两者关系应满足式(3-23)。通常尺寸公差等级在 IT5～IT8 之间时,$K = 25\% \sim 65\%$。

$$T_f = KT_s \tag{3-23}$$

The properties of the workpiece structure: when meeting the function requirements, the geometrical tolerance grades can be decreased 1~2 grade for the workpiece of which structure is complex, poorly rigid (such as thin and long spindles, thin wall workpieces) or not easy to machined and measured, such as: the hole or shaft of which the rate of length to diameter is large; the hole or shaft of which the distance is large; the plane surface of workpiece of which the width is large, usually the width is 1/2 times great than length; the perpendicular or parallelism of the line to line and line to plane related to the plane to plane; the workpiece should follow the specified standards which have been standarded, for example, the cylindricity tolerances of shafts and holes on house fitted with rolling bearing, the straightness tolerance of the railway guides in the machine tools, the parallelism of holes in the gear boxes.

工件结构特点:对于结构复杂、刚性差(如细长轴、薄壁件)或不易加工和测量的工件,与轴有关的孔,在满足功能要求的前提下,几何公差等级可降低 1～2 级。例如:长径比大的孔或轴;距离大的孔或轴;宽度较大的工件的平面,通常宽度是长度的 1/2 倍;线对线和线与面之间的垂直或平行;装有滚动轴承的轴孔的圆柱度、机床导轨的直线度、齿轮箱体的平行度。

3.8.3　General Geometrical Tolerances 未注几何公差

The general geometrical tolerances are specified for machined workpieces in *Geometrical tolerancing Geometrical tolerance for features without individual tolerance indications* (GB/T 1184—1996). These general geometrical tolerances are derived from measurements on workpieces formed by metal removal. Workpieces are mainly measured to which the general dimensional tolerances GB/T 1804-m had been applied. Usually, only such features that have geometrical tolerance indication on the drawing are measured. General tolerance refers to the tolerance which is not marked separately and can be guaranteed under the normal processing conditions in the workshop. For sizes with general tolerance, it is not necessary to indicate the limit deviation after the size.

在《形状和位置公差　未注公差值》(GB/T 1184—1996)中规定了一般几何公差。一般几何公差确定的依据是对金属加工表面的工件测量结果,属于经验积累结果,而被测量工件的尺寸公差为未注尺寸公差(GB/T 1804-m)。通常只测量在图纸上标注几何公差的特征项目。未注公差是指未单独标注出的公差,指在车间通常加工条件下可保证的公差。采用未注公差的尺寸在该尺寸后不需要注出其极限偏差数值。

The general tolerances on straightness and flatness are given in Tab. 3-8. When selecting a straightness tolerance from the table, the length of the corresponding line must be taken, and in the case of a flatness tolerance, the longer length of the surface or the diameter of the circular surface must be taken.

直线度和平直度的未注公差见表 3-8。从表中选择直线度公差时,必须取相应直线的长度为参考量。如果是平面度公差,则必须取较长的表面长度或圆面的直径。

General Tolerances on Straightness and Flatness Tab. 3-8
直线度和平面度的未注几何公差值 表 3-8

Accuracy grades 精度等级	Ranges of nominal length 公称长度范围(mm)					
	~10	>10~30	>30~100	>100~300	>300~1000	>1000~3000
H	0.02	0.05	0.1	0.2	0.3	0.4
K	0.05	0.1	0.2	0.4	0.6	0.8
L	0.1	0.2	0.4	0.8	1.2	1.6

General tolerances on circularity have been established equal to the numerical value of diameter tolerance, or to the respective value of the general tolerance on circular run-out, whichever is the smaller. The circularity deviation can not be larger than the radial circular run-out deviation of the same feature (for geometrical reasons). Therefore, the numerical values of the general tolerances on circular run-out have been taken as the upper limits on the general tolerances on circularity. Whether the roundness deviation may take advantage of the full tolerance zone depends on the diameter tolerance and on the shape of the roundness deviation. In the case of an elliptic form the deviation may only occur up to half of the numerical value of the size tolerance, otherwise actual local sizes would exceed the size tolerance.

圆度的未注公差等于直径公差值或圆跳动公差值中较小者。圆度公差不能大于同一特征项目的径向圆跳动公差。因此,圆跳动未注公差的数值作为圆度未注公差的上限。圆度公差取决于直径公差和圆度公差的公差带形状。在椭圆形状的情况下,公差最多只能为尺寸公差数值的一半,否则实际局部尺寸将超过尺寸公差。

The general tolerance on cylindricity is not provided. The cylindricity deviation consists of two components: circularity deviation and parallelism deviation of opposite generator line (the latter contains the straightness deviation). Each of these components is controlled by its general tolerance. However, it is unlikely that on one workpiece the two components occur with their maximum permissible values and that they accumulate to the theoretical maximum permissible value of the cylindricity deviation. Besides, not enough measured values are presently available to derive suitable general tolerances on cylindricity. However, since the cylindricity deviation is almost only of importance for cylindrical fits, and since the form deviation is already limited by the envelope requirement or by an individually indicated circularity tolerance, there seem to be no need for the establishment of general tolerances on cylindricity. Therefore, they have usually been omitted.

一般不提供圆柱度的未注公差。因为圆柱度公差由圆度公差和素线平行度公差(后者包含直线度公差)两部分组成。它们中的每一个都受其未注公差的控制。然而,在一个工件上,两个项目的最大允许值不太可能同时出现,也不可能同时累积到圆柱度公差的理论最大允许值。此外,目前还没有足够的测量值来推导合适的圆柱度公差。圆柱度公差只对圆柱配合面起作用,并且由于形状公差已经受到包容要求或单独标注的圆度公差限制,没有必要再建立圆柱度的未注公差。因此,圆柱度通常被省略。

Line profile, surface profile and position: the general tolerances on profile, angularity, position and total run-out are specified, since they are limited by teleranced tolerances or by general size tolerances.

轮廓、角度、位置以及总跳动的未注公差,受到几何公差或未注尺寸公差的限制。

Peneral tolerance on parallelism equal to the larger one of the straightness (or flatness) and the size tolerance.

平行度的未注公差等于直线度(或平面度)和尺寸公差中较大的一个。

The general tolerances on perpendicularity are given in Tab. 3-9. The longer of the two sides forming the rectangular angle is to be taken as the datum. If the nominal lengths of the two sides are equal, either of them may apply as the datum. The definition of the geometric tolerance on perpendicularity is as follows: the zone between two parallel straight lines or planes or cylindrical zone that is perpendicular to the datum and within which the actual line or surface or the shaft or median plane shall remain. This tolerance zone also limits the straightness or flatness deviations and the axial run-out deviations of the sides forming the rectangular angle. Therefore, the general tolerance on perpendicularity should not be smaller than the general tolerances on straightness (and flatness) and on axial run-out.

垂直度的未注公差见表3-9。以形成直角的两侧中较长的一个作为基准,如果两侧的公称长度相等,则可将任意一个作为基准。垂直度几何公差的定义是:两条平行直线或平面之间的区域,或垂直于基准的圆柱区域,在该区域内,实际的直线或表面、轴线或中间平面应处在该区域内。该公差带还限制了直线度或平面度公差以及形成直角的侧面的轴向跳动公差。因此,垂直度的未注公差不应小于直线度(和平面度)和轴向跳动的未注公差。

General Tolerances on Perpendicularity　　　　　　　　　　Tab. 3-9
垂直度的未注几何公差值　　　　　　　　　　　　　　　　表3-9

Accuracy grades 精度等级	Ranges of nominal length 公称长度范围(mm)			
	~100	>100~300	>300~1000	>1000~3000
H	0.2	0.3	0.4	0.5
K	0.4	0.6	0.8	1.0
L	0.6	1.0	1.5	2

General tolerances on angularity in the definition is not specified. For angles indicated in theoretically exact sizes the angularity tolerance has to be indicated on the drawing individually.

倾斜度的未注公差并未给出明确定义。对于倾斜度,必须在图纸上单独注明角度的理论正确值。

General tolerance on coaxiality is not specified, and it can be limited by general tolerances on radial circularity run-out.

同轴度,未规定同轴度的未注公差,可由径向圆度跳动的未注公差限制。

The general tolerances on symmetry are given in Tab. 3-10. They apply to symmetrical features. If one of the two features is symmetrical and the other cylindrical, the longer feature is to be taken as the datum. If the nominal lengths of the two features are equal, either of them may apply as the datum. In deriving the values of Tab. 3-10 the following has been observed. As the tolerance zone on symmetry also limits certain straightness or flatness deviations, the general tolerances on symmetry should not be smaller than the general tolerances on straightness and flatness. Furthermore, measurements on workpieces (as described above) revealed that symmetry deviations up to 0.5mm occur independently of the feature length. The reason for this is probably that the general tolerances are set up depending on the largest measured deviations, which were not due to the inaccuracy of the machine tool but rather to inaccuracy when adjusting the workpiece in the machine tool after the workpiece has been turned over (re-chucking). Small and large workpieces were adjusted with the same inaccuracy, and showed the same distribution of the measured symmetry deviations.

对称度的未注公差见表3-10。它们适用于对称性特征。如果两个特征中的一个是对称的,另一个是圆形的,则以较长的特征作为基准。如果两个特征的公称长度相等,则它们中的任何一个都可以作为基准。在推导表3-10中的数值时,应注意到以下几点。对称性公差带限制了某些直线度或平面度误差,一般对称度公差不应小于平面度公差。此外,对工件的测量表明,对称性度误差高达0.5mm,且与特征长度无关。因为未注公差是根据测量出的最大加工误差设置的,这些误差不是由于机床本身产生的,而是由于工件翻转(重新夹紧)后在机床上调整工件时而产生的。若小尺寸和大尺寸工件的调整精度相同,则其测量的对称度误差分布相同。

General Tolerances on Symmetry   Tab.3-10
对称度的未注几何公差值   表3-10

Accuracy grades 精度等级	Ranges of nominal length 公称长度范围(mm)			
	~100	>100~300	>300~1000	>1000~3000
H	0.5			
K	0.6		0.8	1.0
L	0.6	1.0	1.5	2

The general tolerances on circular run-out (radial, axial and inclined circular run-out) are given in Tab.3-11. The bearing surfaces are to be taken as the datum if they are designated as bearing surfaces. In the other case, for radial circular run-out, the longer feature is to be taken as the datum. If the nominal lengths of the two features are equal, either of them may apply as the datum.

圆跳动(径向、轴向和倾斜圆跳动)的未注公差见表3-11。如果支承面被指定,则以支承面为基准。对于径向圆跳动,以较长的特征作为基准。如果两个特性的公称长度相等,则它们中的任何一个都可以作为基准。

General Tolerances on Circular Run-out   Tab.3-11
圆跳动的未注几何公差值   表3-11

Accuracy grades 精度等级	Run-out tolerance 公差值	Precision grades 精度等级	Run-out tolerance 公差值
H	0.1	L	0.5
K	0.2		

The radial total run-out deviation consists of two components: circular run-out deviation and parallelism deviation (the latter contains the straightness deviation). The axial total run-out deviation consists of two components: circular run-out deviation and flatness deviation. General tolerances on radial total run-out are not intended to be standardized, for similar reasons as for cylindricity. General tolerances on axial total run-out are not intended to be standardized, since general tolerances on perpendicularity are already standardized.

径向全跳动由两部分组成:圆跳动和平行度(后者包含直线度)。轴向全跳动由圆跳动和平面度两部分组成。由于与圆柱度公差相似,径向全跳动的未注公差没有明确规定。轴向全跳动的未注公差也未明确,可以参考垂直度的未注公差。

For general tolerances of orientation, location and run-out, it is necessary to determine the datum without drawing indications. The longer of the two considered features applies as the datum. When they are of equal nominal length, either may serve as a datum. An exception applies with general tolerances of run-out when there are bearing surfaces designated as such in the drawing. Then these surfaces serve as the datum(s). Although, with the exception of designated bearing surfaces, datum for general geometrical tolerances is not designated in the drawing, there is no accumulation of general geometrical tolerances possible from one feature to the next etc., because the general geometrical tolerances apply to all possible combinations of any two features of the workpiece. Therefore the datum is not inspected and individual tolerancing should be used.

对于方向、位置和跳动的未注公差,在没有图纸标识的情况下确定基准,两个参考特征中较长的一个用作基准,具有相同的公称长度时,两者都可以作为基准。当图纸中有指定的支撑面时,用这些表面作为基准,跳动公差除外。除指定的支承面外,图纸中未指定未注几何公差的基准,从一个特征传递到下一个特征,未注几何公差不会累积,因为一般的几何公差适用于工件任何两个特征的所有可能的组合。基准不检验,应使用单独的公差。

Exercises 3 习题 3

3-1 How many geometric tolerance features are there? What are their names and symbols?

3-1 几何公差特征共有几项?其名称和符号是什么?

3-2 What are the contents of tolerance principle?

3-2 公差原则包含哪些内容?

3-3 What is the maximum material virtual size?

3-3 什么是最大实体实效尺寸?

3-4 What is the difference between geometric tolerance zone and dimensional tolerance zone? What are the four features of geometric tolerance?

3-4 几何公差带与尺寸公差带有何区别?几何公差的四要素是什么?

3-5 When should there be a ϕ before tolerance value?

3-5 什么时候在公差值前面加符号 ϕ?

3-6 Correct the indication mistakes in Fig. 3-43。

3-6 改正图 3-43 中的错误。

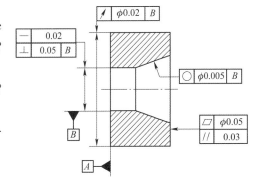

Fig. 3-43 Figure for Exercise 3-6
图 3-43 习题 3-6 图

Chapter 4　Surface Roughness 表面粗糙度

4.1　Basic Concept of Surface Roughness 表面粗糙度的基本概念

4.1.1　Definition of Surface Roughness 表面粗糙度的定义

Surface roughness is the micro geometric characteristics composed of small spacing and peak valley on the machined surface, also known as micro roughness. The formation of surface roughness is mainly due to the friction between surfaces of tools and parts, plastic deformation and metal tearing during chip separation, and high frequency vibration in the processing system. After the part is finished, its profile shape is complex, as shown in Fig. 4-1. After the measurement filter, its transmission characteristic curve is shown in Fig. 4-2.

表面粗糙度是指加工表面上具有的较小间距和峰谷所组成的微观几何形状特性,亦称微观不平度。表面粗糙度主要是由于在加工过程中刀具和零件表面之间的摩擦,切屑分离时的塑性变形和金属撕裂,以及在工艺系统中存在高频振动等原因所形成。零件完工后它的截面轮廓形状是复杂的,如图 4-1 所示。经过测量滤波器后,它的传输特性曲线如图 4-2 所示。

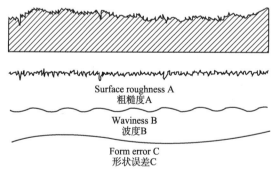

Fig. 4-1　Surface Roughness, Waviness and Form Error on Part Surface
图 4-1　零件表面上的表面粗糙度、波度和形状误差

The profile with wavelength greater than λ_s is called the original actual profile (P profile); the wavelength between λ_s and λ_c is called the surface roughness (R profile); the wavelength between λ_c and λ_f is called the surface waviness (W profile). The profile with wavelength greater than λ_f is the form error. Generally, the wavelength of surface roughness is less than 1mm, and the wavelength of 1~10mm belongs to surface waviness; the wavelength greater than 10mm belongs to macro form error.

Chapter 4 Surface Roughness 表面粗糙度

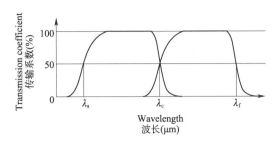

Fig. 4-2 Transmission Characteristics of Surface Roughness and Waviness

图 4-2 表面粗糙度与波纹度的传输特性

波长大于 λ_s 的轮廓称为原始实际轮廓(P 轮廓);波长在 λ_s 和 λ_c 之间称为表面粗糙度(R 轮廓);波长在 λ_c 和 λ_f 之间称为表面波度(W 轮廓)。波长大于 λ_f 就是形状误差。一般表面粗糙度的波长小于 1mm,波长为 1~10mm 的属于表面波度;波长大于 10mm 的属于宏观形状误差。

4.1.2 Influence of Surface Roughness on Mechanical Parts Performance 表面粗糙度对机械零件使用性能的影响

The size of surface roughness has a great influence on the service performance and service life of parts, especially the mechanical parts working under high temperature, high speed, high pressure. In order to select the parameters and allowable values of surface roughness reasonably, it is necessary to understand its influence on performance.

表面粗糙度的大小对零件的使用性能和使用寿命有很大影响,尤其对高温、高速、高压条件下工作的机械零件影响更大。为了合理地选用表面粗糙度的参数及允许值,首先要了解它对使用性能的影响。

4.1.2.1 Influence on Friction and Wear of Moving Surface of Parts 对零件运动表面的摩擦和磨损的影响

The two surfaces with micro geometric errors can only contact at the peak of the contour. The actual effective contact area is very small, which leads to the increase of unit pressure. If there is relative movement between the surfaces, the contact between the two peaks will produce friction resistance to the movement and cause wear of the parts. Generally speaking, the more rough the surface is, the greater the resistance is and the faster the part wears. It must be pointed out that a smoother surface does not necessarily create less wear on the surface. In addition to the influence of surface roughness, the wear loss is also related to the scratching effect of worn metal particles, extrusion of lubricating oil and adsorption between molecules etc. As a result, the wear of particularly smooth surfaces is aggravated.

具有微观几何形状误差的两个表面只能在轮廓的峰顶发生接触,实际有效接触面积很小,导致单位压力增大。若表面间有相对运动,则峰顶间的接触作用就会对运动产生摩擦阻力,同时使零件产生磨损。一般来说,表面越粗糙,则摩擦阻力越大,零件的磨损也越快。必须指出,表面越光滑,磨损量不一定越小。磨损量除受表面粗糙不平的影响外,还与磨损下来的金属微粒的刻划作用、润滑油被挤出和分子间的吸附作用等因素有关。所以,特别光滑的表面磨损反而加剧。

4.1.2.2　Influence on Fits Properties 对配合性质的影响

For the parts with fitting requirements, the surface roughness affects the stability of the fit properties of no matter what kind of fit. For the clearance fit of sliding bearing, the gap will increase due to the wear of peak tip with surface micro shape in the working process. If the gap caused by the rougher surface increases too much, the original fitting properties will be destroyed. The surface roughness is not related to the nominal size, so the smaller the fit size is, the more serious the effect is. For the transition fit, the rough surface will also enlarge the gap in the process of use and assembly, thus reducing the degree of centering and changing the original fitting properties. For interference fit, due to the uneven surface of the parts, the peak of part surface will be flattened after pressing, so that the actual interference is less than the theoretical interference, thus reducing the connection strength. Therefore, in order to ensure the parts fitting properties, grinding method is usually applied to improve the parts surface roughness.

对于有配合要求的零件表面,无论是哪一类配合,表面粗糙度都影响配合性质的稳定性。对于滑动轴承用的间隙配合,会因表面微观形状的峰尖在工作过程中很快被磨损而使间隙增大。如果表面粗糙所引起的间隙增大过多,就会破坏原有的配合性质。由于表面粗糙度与公称尺寸的大小无关,所以配合的尺寸越小,这种影响越严重。对于过渡配合,表面粗糙也会在使用和装拆中过程中使间隙扩大,从而降低定心程度,改变原来的配合性质。对于过盈配合,由于零件表面凸凹不平,配合零件经过压装后,零件表面的峰顶会被挤平,以致实际过盈小于理论的计算过盈量,从而降低了连接强度。因此,为了保证零件的配合性质,通常采用磨削的方法提高零件的表面粗糙度质量。

4.1.2.3　Influence on Corrosion Resistance 对抗腐蚀性的影响

Metal corrosion is often caused by chemical or electrochemical action, such as steel rust, copper green. The rougher the surface is, the more serious the corrosion is. Since corrosive gas or liquid is easy to accumulate at the bottom of the pit, corrosion will penetrate into the metal from the valley. The rougher the surface is and the deeper the pit is, the more serious the corrosion is. Therefore, improving the parts surface roughness can enhance its corrosion resistance.

金属腐蚀往往是由于化学作用或电化学作用造成的,如钢铁生锈、铜生铜绿,都是腐蚀作用所致。零件表面越粗糙,其腐蚀作用也就越强。由于腐蚀性气体或液体容易积存在凹谷底,腐蚀作用便从凹谷深入到金属内部去。表面越粗糙,凹谷越深,腐蚀作用就越严重。因此提高零件表面粗糙度质量,可以增强其抗腐蚀能力。

4.1.2.4　Influence of Anti Fatigue Strength 对抗疲劳强度的影响

Under the conditions of alternating load, heavy load and high-speed working, the parts fatigue strength is related to the physical and mechanical properties of the material, and also to the surface roughness. Because the rougher the surface is and the deeper the dent is, the smaller the root curvature radius is, and the more sensitive it is to stress concentration. Especially under the action of alternating load, the influence is greater, so the parts will quickly produce fatigue cracks and damage. Therefore, for the parts under alternating load, the fatigue strength can be improved by improving the surface roughness quality, thus the size and weight of the parts can be reduced accordingly.

零件在交变载荷、重载荷及高速工作条件下,其疲劳强度除了与零件材料的物理、力学性能有关外,还与表面粗糙度有很大关系。因为零件表面越粗糙,凹痕越深,其根部曲率半径就越小,对应力集中就越敏感。特别是在交变负荷作用下,其影响更大,零件往往因此而很快产生疲劳裂缝而损坏。所以对于承受交变载荷的零件,若提高其表面粗糙度质量,则可提高其疲劳强度,从而可以相应减小零件的尺寸和重量。

4.1.2.5　Influence on Contact Sealing 对结合密封性的影响

If the two rough joint surfaces only contact at local points, there will be gaps, which will affect the sealing performance. For the static sealing surface without relative sliding between the contact surfaces, if the bottom of the surface micro unevenness is too deep, and the sealing material can not completely fill these micro uneven bottoms after the assembly is preloaded, the leakage gap will be left on the sealing surface. Therefore, improving the surface roughness quality of parts can improve its sealing performance. For the relatively sliding dynamic sealing surface, due to the relative movement, there should be a certain thickness of lubricating oil film between the surfaces, so the micro uneven height of the surface should be appropriate, which is generally $4 \sim 5 \mu m$.

粗糙不平的两个结合表面,仅在局部点上接触必然产生缝隙,影响密封性能。对于接触表面之间没有相对滑动的静力密封表面,若表面微观不平度谷底过深,则密封材料在装配受预压后还不能完全填满这些微观不平谷底,从而在密封面上留下渗漏间隙。因此,提高零件表面粗糙度质量,可提高其密封性能。对于相对滑动的动力密封表面,由于相对运动,表面间需要有一定厚度的润滑油膜,所以表面微观不平高度应适宜,一般为 $4 \sim 5 \mu m$。

4.1.2.6　Other Influences 其他影响

The influence of surface roughness on parts performance is far more than these five aspects, such as the impact on contact stiffness, impact strength, resistance of fluid flow, surface high frequency current, appearance quality and measurement accuracy of machines and instruments.

表面粗糙度对零件性能的影响远不止这5个方面。如对接触刚度的影响,对冲击强度的影响,对流体流动的阻力的影响,对表面高频电流的影响,以及对机器、仪器的外观质量和测量精度等都有很大影响。

In a word, surface roughness is an important parameter in precision design. In order to ensure mechanical parts performance, the requirements of surface roughness must be put forward reasonably.

总之,表面粗糙度是精度设计中的一个重要的参数。为保证机械零件的使用性能,必须合理地提出表面粗糙度要求。

4.2　Evaluation of Surface Roughness 表面粗糙度的评定

For parts with surface roughness requirements, it is necessary to measure and evaluate the qualification of surface roughness after machining.

具有表面粗糙度要求的零件表面,加工后需要测量和评定其表面粗糙度的合格性。

4.2.1 Basic Terms 基本术语

4.2.1.1 Sampling length 取样长度(lr)

The sampling length refers to the length in the x-axis direction used to distinguish the irregular features of the evaluated profile. It is the length of a datum line specified in the measurement or evaluation of surface roughness, which contains at least five profile peaks and profile valleys. The sampling for evaluating the surface roughness is obtained from the surface profile. The direction of the sampling length lr is consistent with the direction of the profile, that is, the direction of the x-axis is consistent with the direction of the spacing. The purpose of specifying the sampling length is to limit and weaken the influence of other geometric errors, especially the surface waviness, on the measurement. Generally, the rougher the surface is, the larger the sampling length is.

取样长度是指用于判别被评定轮廓的不规则特征的 x 轴方向上的长度,是测量或评定表面粗糙度时所规定的一段基准线长度,它包含至少 5 个轮廓峰和轮廓谷。评定表面粗糙度的取样是从表面轮廓上取得的,取样长度 lr 的方向与轮廓走向一致,即 x 轴方向与间距方向一致。规定取样长度的目的在于限制和减弱其他几何形状误差,特别是表面波度对测量的影响。一般来说,表面越粗糙,取样长度就越大。

4.2.1.2 Evaluation length 评定长度(ln)

The evaluation length is used to judge the length in the x-axis direction of the evaluated contour. Due to the uneven surface roughness of parts, in order to reflect the characteristics of surface roughness reasonably, the minimum length specified in the measurement and evaluation of surface roughness is called evaluation length (ln). The evaluation length may include one or more sampling lengths. Generally, $ln = 5lr$; if the measured surface is relatively uniform, select $ln < 5lr$; if the uniformity is poor, choose $ln > 5lr$.

评定长度是用于判别被评定轮廓的 x 轴方向上的长度。由于零件表面粗糙度不均匀,为了合理地反映表面粗糙度特征,在测量和评定表面粗糙度时所规定的一段最小长度称为评定长度(ln)。评定长度可包含一个或几个取样长度。一般情况下,取 $ln = 5lr$;若被测表面比较均匀,可选 $ln < 5lr$;若均匀性差,可选 $ln > 5lr$。

4.2.1.3 Median Line 中线

The median line is the datum line which has the geometric contour shape and divides the contour. There are two kinds of datum lines: the least square median line of the contour and the arithmetic mean median line of the contour. The least squares median line of contour is the line that makes the offset distance of each point on the contour line the minimum, that is, the sum of squares of ordinate values, within the sampling length. The arithmetic mean median line of contour is a line that divides the actual contour into upper and lower parts within the sampling length and makes the area of the upper and lower parts equal. It is difficult to determine the position of the least squares median line on the contour graph. The arithmetic mean median line of the contour can be used, and the arithmetic mean median line is usually determined by visual estimation.

中线是具有几何轮廓形状并划分轮廓的基准线。基准线有轮廓最小二乘中线和轮廓算术平均中线两种。轮廓最小二乘中线是指在取样长度内,使轮廓线上各点偏距最小,也就是纵坐标值的平方和为最小的线。轮廓算术平均中线是指在取样长度内,将实际轮廓划分为上、下两部分,且使上下两部分面积相等的线。在轮廓图形上确定最小二乘中线的位置比较困难,可使用轮廓算术平均中线,通常用目测估计确定算术平均中线。

4.2.2 Evaluation Parameters 评定参数

In order to meet the different functional requirements for the surface of parts, *Geometrical product specifications (GPS)—Surface texture: Profile method—Terms, definitions and surface texture parameters* (GB/T 3505—2009) specifies the corresponding evaluation parameters from three aspects of surface micro geometry amplitude, spacing and shape.

为了满足对零件表面不同的功能要求,《产品几何技术规范(GPS) 表面结构 轮廓法 术语、定义及表面结构参数》(GB/T 3505—2009)从表面微观几何形状幅度、间距和形状3个方面的特征,规定了相应的评定参数。

4.2.2.1 Amplitude Parameter 幅度参数

The arithmetic mean deviation Ra of the contour is the arithmetic mean of the absolute value of the ordinate value within a sampling length. The value of Ra can objectively reflect the micro geometric characteristics of the measured surface. The smaller the Ra value is, the smaller the amplitude of the micro peaks and valleys on the measured surface is, the smoother the surface is; conversely, the larger the Ra value, the rougher the surface is. Due to the limitation of the radius of the stylus and the measuring principle of the instrument, the Ra value should not be used as the evaluation parameter of the too rough or too smooth surface, but only suitable for the surface with Ra value of $0.025 \sim 6.3 \mu m$.

轮廓的算术平均偏差 Ra 是在一个取样长度内纵坐标值绝对值的算术平均值。Ra 值的大小能客观地反映被测表面微观几何特性,Ra 值越小,说明被测表面微小峰谷的幅度越小,表面越光滑;反之,Ra 越大,说明被测表面越粗糙。Ra 值是用触针式电感轮廓仪测得的,受触针半径和仪器测量原理的限制,不宜用作过于粗糙或太光滑表面的评定参数,仅适用于 Ra 值为 $0.025 \sim 6.3 \mu m$ 的表面。

The maximum height Rz of the contour is the height of the sum of the maximum contour peak height and the maximum contour Valley depth within a sampling length. The peak of surface roughness profile refers to the part of surface roughness profile connecting two adjacent intersection points outward; the surface roughness profile Valley refers to the part of roughness profile connecting two adjacent intersection points inward.

轮廓的最大高度 Rz 是在一个取样长度内,最大轮廓峰高和最大轮廓谷深之和。表面粗糙度轮廓峰是指连接表面粗糙度轮廓和横坐标轴相交的两相邻交点向材料外的表面粗糙度轮廓部分;表面粗糙度轮廓谷是指连接表面粗糙度轮廓和横坐标轴相交的两相邻交点向材料内的粗糙度轮廓部分。

Amplitude parameters (Ra, Rz) are the parameters that must be marked in the standard, so they are also called basic parameters.

幅度参数(Ra、Rz)是标准规定必须标注的参数,故又称基本参数。

4.2.2.2 Spacing Parameter 间距参数

The spacing parameter is expressed by the average width RSm of contour units, which is defined as the average width of contour units within a sampling length.

间距参数用轮廓单元的平均宽度 RSm 来表示,RSm 是在一个取样长度内轮廓单元宽度的平均值。

The outward and inward parts of the evaluation profile at the beginning or the end of the sampling length are regarded as a profile peak or valley of surface roughness. When several surface roughness profile units are determined on several continuous sampling lengths, the peaks and valleys evaluated at the beginning or end of each sampling length are included only once at the beginning of each sampling length.

在取样长度始端或末端的评定轮廓的向外部分和向内部分看作是一个表面粗糙度轮廓峰或轮廓谷。当在若干个连续的取样长度上确定若干个表面粗糙度轮廓单元时,在每一个取样长度的始端或末端评定的峰和谷仅在每个取样长度的始端计入一次。

4.2.2.3 Mixing Parameters 混合参数

The mixing parameter is expressed as the ratio of the solid material length of the contour to the evaluated length at a given horizontal position c.

混合参数用轮廓的支承长度率 $Rmr(c)$ 来表示。轮廓的支承长度率 $Rmr(c)$ 定义为在给定水平位置 c 上轮廓的实体材料长度与评定长度的比率。

The spacing parameter (RSm) and the mixing parameter $Rmr(c)$ are relative to the basic parameters, which are called additional parameters. Only when the important surface of a few parts has special use requirements, the two additional evaluation parameters can be selected. The additional parameters can not be separately indicated on the drawing, but can only be used as auxiliary parameters of amplitude parameters.

间距参数(RSm)与混合参数 $Rmr(c)$,是相对于基本参数而言,它们被称为附加参数。只有在少数零件的重要表面有特殊使用要求时,才选用这两个附加评定参数,附加参数不能单独在图样上注出,只能作为幅度参数的辅助参数注出。

4.3 Parameter Value of Surface Roughness and Its Selection 表面粗糙度的参数值及其选用

4.3.1 Parameter Values of Surface Roughness 表面粗糙度的参数值

The parameter values of surface roughness have been standardized, so the design should be selected according to the national standard *Geometrical product specifications(GPS)—Surface texture: Profile method—Surface roughness parameters and their values* (GB/T 1031—2009). The amplitude parameter values are shown in Tab.4-1 and Tab.4-2, the interval parameter values are shown in Tab.4-3, and the mixing parameter values are shown in Tab.4-4.

表面粗糙度的参数值已经标准化,设计时应按《产品几何技术规范(GPS) 表面结构 轮廓法 表面粗糙度参数及其数值》(GB/T 1031—2009)规定的参数值系列选取。幅度参数值见表4-1和表4-2,间距参数值见表4-3,混合参数值见表4-4。

Parameter Values of Ra — Tab. 4-1
Ra 的数值 — 表4-1

0.012	0.025	0.050	0.100	0.20
0.40	0.80	1.60	3.2	6.3
12.5	25	50	100	—

Parameter Values of Rz — Tab. 4-2
Rz 的数值 — 表4-2

0.025	0.050	0.100	0.20	0.40
0.80	1.60	3.2	6.3	12.5
25	50	100	200	400
800	1600	—	—	—

Parameter Values of RSm — Tab. 4-3
RSm 的数值 — 表4-3

0.006	0.0125	0.025	0.050	0.100
0.20	0.40	0.80	1.6	3.2
6.3	12.5	—	—	—

Parameter Values of $Rmr(c)$ — Tab. 4-4
$Rmr(c)$ 的数值 — 表4-4

10%	15%	20%	25%	30%
40%	50%	60%	70%	80%
90%	—	—	—	—

In general, when measuring Ra and Rz, it is recommended to select the corresponding sampling length and evaluation length value according to Tab. 4-5. At this time, the sampling length value can be omitted in the drawing. When the values in Tab. 4-5 cannot be selected for special requirements, the sampling length value shall be marked on the drawing.

在一般情况下,测量 Ra 和 Rz 时,推荐按表4-5选用对应的取样长度及评定长度值,此时在图样上可省略标注取样长度值。当有特殊要求不能选用表4-5中数值时,应在图样上标注出取样长度值。

Parameter Values of lr and ln — Tab. 4-5
lr 和 ln 的参数值 — 表4-5

Ra	Rz	lr	ln
≥0.008~0.02	≥0.025~0.10	0.08	0.4
>0.02~0.10	>0.10~0.50	0.25	1.25
>0.1~2.0	>0.50~10.0	0.8	4.0
>2.0~10.0	>10.0~50.0	2.5	12.5
>10.0~80.0	>50.0~320	8.0	40.0

4.3.2 Selection of Surface Roughness 表面粗糙度的选用

4.3.2.1 Selection of Evaluation Parameters 评定参数的选用

Amplitude parameters are the basic parameters specified in the standard and can be selected independently. For the surface with surface roughness requirements, an amplitude parameter must be selected. For the surface roughness parameter value of $0.025 \sim 6.3 \mu m$ in amplitude direction, Ra is recommended to be preferred, because Ra can reflect the micro peak valley characteristics of the measured surface. When $Ra < 0.025 \mu m$ or $Ra > 6.3 \mu m$, Rz should be selected.

幅度参数是标准规定的基本参数,可以独立选用。对于有表面粗糙度要求的表面,必须选用一个幅度参数。对于幅度方向的表面粗糙度参数值为 $0.025 \sim 6.3 \mu m$ 的零件表面,标准推荐优先选用 Ra,因为 Ra 能够比较全面地反映被测表面的微小峰谷特征。在表面粗糙度要求特别高或特别低,$Ra < 0.025 \mu m$ 或 $Ra > 6.3 \mu m$ 时,选用 Rz。

Only a few additional parameters can not be used as additional parameters for general evaluation. RSm is mainly used when there are requirements for painting performance, such as uniform spraying, excellent adhesion and smoothness of coating. In addition, when it is required to resist crack, vibration, corrosion and reduce the friction resistance of fluid flow, it is also selected.

附加评定参数 RSm 一般不能作为独立参数选用,只有少数零件的重要表面,有特殊使用要求时才附加选用。RSm 主要在对涂漆性能,如喷涂均匀、涂层有极好的附着性和光洁性等有要求时选用。另外,要求冲压成形时抗裂纹、抗振、抗腐蚀、减小流体流动摩擦阻力等时也选用 RSm。

For the selection of mixed parameters, the support length ratio $Rmr(c)$ of the contour is mainly used in the case of high requirements of wear resistance and contact stiffness.

对于混合参数的选用,轮廓的支承长度率 $Rmr(c)$ 主要在耐磨性、接触刚度要求较高等场合附加选用。

4.3.2.2 Selection of Parameter Values 参数值的选用

The reasonable choice of surface roughness parameters not only has a great impact on the performance of products, but also has a direct impact on the quality and manufacturing cost of products. Generally speaking, the smaller the surface roughness value (evaluation parameter value) is, the better the working performance and the longer the service life of the parts is. The selection of surface roughness parameters should consider not only the functional requirements of parts, but also the manufacturing cost. The selection principle of surface roughness parameters is to meet the functional requirements of parts, and the second is to consider the possibility of economy and technology.

表面粗糙度参数值选择的合理与否,不仅对产品的使用性能有很大的影响,而且直接关系到产品的质量和制造成本。一般来说,表面粗糙度值(评定参数值)越小,零件的工作性能越好,使用寿命也越长。选择表面粗糙度参数值既要考虑零件的功能要求,又要考虑其制造成本。表面粗糙度参数值的选用原则是首先满足零件的功能要求,其次考虑经济性及工艺的可能性。

The following principles should be followed when selecting parameter values. On the same part, the Ra or Rz value of working surface is smaller than that of non working surface. The value of Ra or Rz of friction surface is smaller than that of non friction surface; the roughness parameter of rolling friction surface is smaller than that of sliding friction surface. The surface roughness of fillet groove of important parts with high velocity, high pressure per unit area and alternating stress should be smaller. The surface roughness values of the fiting surfaces with high fitting properties (such as those with small clearance fit) and interference fit surfaces with reliable connection and heavy load should be smaller; the surface roughness of clearance fit should be smaller than that of interference fit. For the same fit property, the smaller the part size is, the smaller the surface roughness parameter value is; for the same tolerance grade, the surface roughness parameter value of small size is smaller than that of large size, and that of shaft is smaller than that of hole. When determining the parameter value of surface roughness, it should be noted that it is compatible with dimensional tolerance and geometric tolerance. The smaller the dimensional tolerance and geometric tolerance is, the smaller the Ra or Rz value of surface roughness is. The surface roughness value required for corrosion-resistant and good sealing performance or beautiful appearance should be small. If the relevant standards have specified the surface roughness requirements, the surface roughness parameters should be determined according to the standards.

选择表面粗糙度参数值时遵循如下原则。同一零件上,工作表面的 Ra 或 Rz 值比非工作表面小。摩擦表面 Ra 或 Rz 值比非摩擦表面小;滚动摩擦表面比滑动摩擦表面的粗糙度参数值要小。运动速度高、单位面积压力大,以及受交变应力作用的重要零件的圆角沟槽的表面粗糙度值都应较小。配合性质要求高的配合表面(如小间隙配合的配合表面)以及要求连接可靠、受重载荷作用的过盈配合表面的表面粗糙度值都应较小;间隙配合比过盈配合的表面粗糙度值要小。配合性质相同,零件尺寸越小则表面粗糙度参数值应越小;同一公差等级,小尺寸比大尺寸、轴比孔的表面粗糙度参数值要小。在确定表面粗糙度参数值时,应注意它与尺寸公差和几何公差协调。尺寸公差和几何公差值越小,表面粗糙度的 Ra 或 Rz 值应越小。要求防腐蚀、密封性能好或外表美观的表面粗糙度数值应较小。凡有关标准已对表面粗糙度要求作出规定,则应按该标准确定表面粗糙度参数值。

In engineering practice, because the relationship between surface roughness and function is very complex, it is difficult to accurately determine the allowable value of parameters. In the specific design, except for the surface with special requirements, the empirical statistical data are generally used and the class comparison method is used. After the surface roughness is preliminarily determined according to the analogy method, the appropriate adjustment is made according to the working conditions.

在工程实践中,由于表面粗糙度和功能的关系十分复杂,因而很难准确地确定参数的允许值。在具体设计时,除有特殊要求的表面外,一般采用经验统计资料,用类比法来选用。用类比法初步确定表面粗糙度后,再对比工作条件做适当调整。

4.4 Surface Roughness Symbol and Its Marking
表面粗糙度符号及其标注

The surface roughness symbols and codes marked on the drawings are the requirements after the surface is completed. The marking of surface roughness shall comply with the national standard *Geometrical Product Specifications (GPS)—Indication of surface texture in technical Product documentation*(GB/T 131—2006).

图样上所标注的表面粗糙度符号和代号是该表面完工后的要求。表面粗糙度的标注应符合国家标准《产品几何技术规范(GPS) 技术产品文件中表面结构的表示法》(GB/T 131—2006)的规定。

4.4.1 Basic Symbols of Surface Roughness 表面粗糙度的基本符号

Tab. 4-6 shows the basic symbols and descriptions of the surface roughness of parts on the drawing. When only machining is required (by material removal or without material removal) but there is no requirement for other provisions on surface roughness, it is allowed to note only the surface roughness symbol.

图样上表示零件表面粗糙度的基本符号及其说明见表 4-6。当仅需要加工(采用去除材料的方法或不去除材料的方法)但对表面粗糙度的其他规定没有要求时,允许只注表面粗糙度符号。

Symbols of Surface Roughness Tab. 4-6
表面粗糙度符号 表 4-6

Symbol 符号	Significance and explanation 意义及说明
∨	Basic symbol indicating that the surface can be obtained by any method 基本符号,表示表面可用任何方法获得
∇	The basic symbol plus a dash indicates that the surface is obtained by removing the material. For example: turning, milling, drilling, grinding, etc. 基本符号加 1 段短划线,表示表面是用去除材料的方法获得。例如:车削、铣削、钻削、磨削等
∨°	The basic symbol plus a small circle indicates that the surface is obtained without removing the material. For example: casting, forging, stamping, etc. 基本符号加 1 个小圆,表示表面是用不去除材料的方法获得。例如:铸造、锻造、冲压成形等
✓ ✓ ✓	A horizontal line can be added on the long side of the above three symbols to mark the relevant parameters and instructions 在前述 3 个符号的长边上均可加 1 个横线,用于标注有关参数和说明
✓° ✓° ✓°	All surfaces with the same surface roughness can be indicated by adding one small circle to the above symbols 在前述 3 个符号的上均可加 1 个小圆,表示所有表面具有相同的表面粗糙度要求

Chapter 4　Surface Roughness 表面粗糙度

4.4.2 Marking of Surface Roughness Symbols 表面粗糙度符号的标注

4.4.2.1 Symbols of Surface Roughness 表面粗糙度的符号

Around the surface roughness symbol, it is required to write some values and relevant regulations. The positions of these values and relevant regulations are shown in Fig. 4-3. The surface roughness symbol, these values and various relevant regulations constitute the surface roughness code. It is stipulated in the standard that when the number of allowable measured values of surface roughness parameters is less than 16% of the total, the upper or lower limit value of surface roughness parameters shall be marked on the drawing, that is, "the 16% rule". When it is required that all measured values of surface roughness parameters shall not exceed the specified values, the maximum value of surface roughness parameters, i.e. "the maximum rule", shall be marked on the drawing.

在表面粗糙度符号周围,要求注写若干数值以及有关规定,这些数值和有关规定注写位置如图 4-3 所示。表面粗糙度符号和这些数值以及各种有关规定共同组成表面粗糙度代号。国标中规定,当允许表面粗糙度参数的所有实测值中超过规定值的个数少于总数的 16%时,应在图样上标注表面粗糙度参数的上限值或下限值,即"16% 规则"。当要求在表面粗糙度参数的所有实测值中不得超过规定值时,应在图样上标注表面粗糙度参数的最大值,即"最大规则"。

In Fig. 4-3, a is the first surface roughness requirement (unit: μm); b is the second surface roughness requirement (unit: μm); c is the processing method (turning, milling, etc.); d is the processing texture direction symbol; e is the machining allowance (unit: mm).

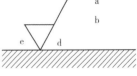

Fig. 4-3　Marking Method of Surface Roughness Code

图 4-3 中,a 为第一表面粗糙度要求(单位为 μm);b 为第二表面粗糙度要求(单位为 μm);c 为加工方法(车,铣等);d 为加工纹理方向符号;e 为加工余量(单位为 mm)。

图 4-3　表面粗糙度代号的注法

4.4.2.2 Marking of Surface Roughness Amplitude Parameters 表面粗糙度幅度参数的标注

The surface roughness amplitude parameters are marked at the positions of codes a and b. the complete graphic symbols are shown in Fig. 4-4. The figure includes marking of upper or lower limit: upper limit symbol "U" and lower limit symbol "L" shall be marked when indicating bidirectional limit. If the same parameter has two-way limit requirement, the label of "U" and "L" can be omitted without ambiguity. If it is a one-way lower limit value, it is necessary to add "L".

表面粗糙度幅度参数标注在代号 a 和 b 位置,完整图形符号如图 4-4 所示。图中包含上限或下限的标注:表示双向极限时应标注上限符号"U"和下限符号"L"。如果同一参数具有双向极限要求,在不引起歧义时,可省略"U"和"L"的标注。若为单向下限值,则必须加注"L"。

Marking of transmission band and sampling length: transmission band refers to the wavelength range between cut-off wavelength values of two filters. The cut-off wavelength of the long wave filter is the sampling length ln. When marking the transmission belt, the short wave is in the front and the long wave is in the back, separated by a hyphen " - ", as 0.0025 - 0.8 shown in figure.

In some cases, only one filter should be marked in the label of transmission belt, and the hyphen "-" should be reserved to distinguish short waves from long waves.

传输带和取样长度的标注：传输带是指两个轮廓滤波器的截止波长值之间的波长范围。长波滤波器的截止波长值就是取样长度 ln。传输带的标注时，短波在前，长波在后，并用连字号"-"隔开，如图中 0.0025 - 0.8。在某些情况下，传输带的标注中，只标一个滤波器，也应保留连字号"-"，来区别是短波还是长波。

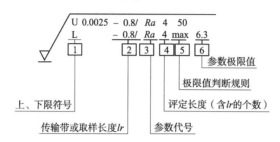

Fig. 4-4 Complete Graphic Symbols of Surface Roughness Amplitude Parameters

图 4-4 表面粗糙度幅度参数完整图形符号

Marking of parameter code: the parameter code is marked after the conveyor belt or sampling length, which is separated by "/", and is marked with Ra or Rz.

参数代号的标注：参数代号标注在传输带或取样长度后，它们之间用"/"隔开，并注明 Ra 或 Rz。

Marking of evaluation length: if the default evaluation length is selected, the annotation can be omitted. If it is not equal to $5lr$, the number of sampling length lr should be noted.

评定长度的标注：如果默认的评定长度时，可省略标注。如果它不等于 $5lr$，则应注明取样长度 lr 的个数。

Limit value judgment rule and limit value annotation: the annotation of limit value judgment rule is shown in Fig. 4-4, the upper limit is "16% Rule", and the lower limit is "maximum rule". To avoid misunderstanding, insert a space between the parameter code and the limit value.

极限值判断规则和极限值的标注：极限值判断规则的标注如图 4-4 中所示上限为"16% 规则"，下限为"最大规则"。为了避免误解，在参数代号和极限值之间插入 1 个空格。

The marking method and significance of surface roughness amplitude parameters are shown in Tab. 4-7.

表面粗糙度幅度参数的标注方法及意义如表 4-7 所示。

Marking Method and Significance of Surface Roughness Amplitude Parameters Tab. 4-7

表面粗糙度幅度参数的标注方法及意义 表 4-7

Symbol 符号	Significance and explanation 意义及说明
$\sqrt{}$ Rz 0.4	Material removal is not allowed, one-way upper limit value, default conveyor belt, maximum height of contour is 0.4μm, evaluation length is 5 sampling lengths (default), 16% rule (default) 不允许去除材料，单向上限值，默认传输带，轮廓的最大高度为 0.4μm，评定长度为 5 个取样长度(默认)，16% 规则(默认)

Chapter 4　Surface Roughness 表面粗糙度

Continued 续上表

Symbol 符号	Significance and explanation 意义及说明
√ Rzmax 0.2	Material removal, one-way upper limit, default conveyor belt, maximum height of contour is 0.2μm, evaluation length is 5 sampling lengths (default), maximum rule 去除材料，单向上限值，默认传输带，轮廓的最大高度为 0.2μm，评定长度为 5 个取样长度（默认），最大规则
√ U Ramax 3.2 　 L Ra 0.8	Material removal is not allowed. The two limit values are the default conveyor belt. The upper limit value: arithmetic mean deviation is 3.2μm, evaluation length is 5 sampling lengths (default), maximum rule; the lower limit value: arithmetic mean deviation is 0.8μm, evaluation length is 5 sampling lengths (default), 16% rule (default) 不允许去除材料，双向极限值，两个极限值均为默认传输带，上限值：算术平均偏差为 3.2μm，评定长度为 5 个取样长度（默认），最大规则；下限值：算术平均偏差为 0.8μm，评定长度为 5 个取样长度（默认），16% 规则（默认）
√ L Ra 1.6	Arbitrary processing method, one-way lower limit, default conveyor belt, arithmetic mean deviation of 1.6μm, evaluation length of 5 sampling lengths (default), 16% rule (default) 任意加工方法，单向下限值，默认传输带，算术平均偏差为 1.6μm，评定长度为 5 个取样长度（默认），16% 规则（默认）
√ 0.008-0.8/Ra 3.2	Remove the material, one-way upper limit value, conveyor belt 0.008~0.8mm, arithmetic mean deviation is 3.2μm, evaluation length is 5 sampling length (default), 16% rule (default) 去除材料，单向上限值，传输带 0.008~0.8mm，算术平均偏差为 3.2μm，评定长度为 5 个取样长度（默认），16% 规则（默认）
Milling 铣 √ Ra 0.8 ⊥ -2.5/Rz 3.2	Remove the material, two one-way upper limit values. The former: default conveyor belt and default evaluation length, arithmetic mean deviation is 0.8μm, 16% rule (default); the latter: conveyor belt is -2.5mm, default evaluation length, maximum height of contour is 3.2μm, 16% rule (default). The surface texture is perpendicular to the projection plane of the view. The processing method is milling 去除材料，两个单向上限值。前者：默认传输带和默认评定长度，算术平均偏差为 0.8μm，16% 规则（默认）；后者：传输带为 -2.5mm，默认评定长度，轮廓的最大高度为 3.2μm，16% 规则（默认）。表面纹理垂直于视图所在的投影面。加工方法为铣削
3√ 0.008-4/Ra 50 　 0.008-4/Ra 6.3	Remove the material, bidirectional limit value. The arithmetic mean deviation of the upper limit value is 50um, the arithmetic mean deviation of the lower limit value is 6.3μm, the upper and lower limit of the conveyor belt is 0.008~4mm, the default evaluation length, 16% rule (default). The machining allowance is 3 mm 去除材料，双向极限值。上限值的算术平均偏差为 50um，下限值的算术平均偏差为 6.3μm，上下极限的传输带均为 0.008~4mm，默认评定长度，16% 规则（默认）。加工余量为 3mm

4.4.2.3 Marking of Other Items of Surface Roughness 表面粗糙度其他项目的标注

The c position of surface roughness code is the location of marking processing method. If the surface roughness of a surface is required to be obtained by specified processing methods (such as polishing, milling, plating, etc.), it can be marked with text.

表面粗糙度代号的 c 位置是标注加工方法的位置。如果某表面的表面粗糙度要求由指定的加工方法(如抛光、铣削、镀覆等)获得,则可以用文字标注。

The d position of surface roughness code is the position of marking the processing texture direction. If it is necessary to control the processing texture direction of the part surface, it can be marked according to the symbol of the processing texture direction.

表面粗糙度代号的 d 位置是标注加工纹理方向的位置。如果需要控制零件表面的加工纹理方向,则可按加工纹理方向的符号进行标注。

The e position of surface roughness code is the position of marking machining allowance, which refers to the total allowance of the part surface before obtaining the surface roughness requirement.

表面粗糙度代号的 e 位置是标注加工余量的位置。加工余量是指获得本表面粗糙度要求前零件表面的总余量。

4.4.3 Example of Surface Roughness Marking 表面粗糙度标注示例

There are two types of surface roughness marking in *Geometrical Product Specifications (GPS)—Indication of surface texture in technical Product documentation* (GB/T 131—2006). One type is to allow the expression of surface roughness requirements in text. The standard stipulates that the words "PAP", "MRR" and "NMR" can be used to indicate that any process is allowed to obtain the surface, the method of removing material is allowed to obtain the surface, and the method of no material removal is allowed to obtain the surface. Another type is to mark on the drawing, as shown in Tab. 4-8.

在《产品几何技术规范(GPS) 技术产品文件中表面结构的表示法》(GB/T 131—2006)中,表面粗糙度标注有两种,一种是允许用文字的方式表达表面粗糙度的要求,标准规定在报告和合同的文本中可以用文字"PAP""MRR"和"NMR"分别表示允许用任何工艺获得表面、允许用去除材料的方法获得表面以及允许用不去除材料方法获得表面。另一种方法是在图样上的标注,如表4-8所示。

Example of Marking Surface Roughness Tab. 4-8
表面粗糙度标注示例 表 4-8

Serial number 序号	Text 在文本中	Pattern 在图样上
1	MRR Ra 0.8; Rz1 3.2	Ra 0.8 Rz1 3.2
2	MRR Ramax 0.8; Rz1max 3.2	Ramax 0.8 Rz1max 3.2

Chapter 4　Surface Roughness 表面粗糙度

Continued 续上表

Serial number 序号	Text 在文本中	Pattern 在图样上
3	MRR U Rz0.8;L Ra0.2	∇ U Rz　0.8 　 L Ra　0.2
4	MRR 车 Rz 3.2	车 ∇ Rz　3.2

4.4.4　Marking Position and Direction of Surface Roughness Symbol 表面粗糙度符号的标注位置与方向

Surface roughness requires that each surface should be marked only once, and should be marked on the same view of corresponding size and tolerance as far as possible. Unless specified, the marked surface roughness requirements are for the finished workpiece surfaces.

表面粗糙度要求对每一表面一般只标注一次,并尽可能标注在相应的尺寸及其公差的同一视图上。除另有说明外,所标注的表面粗糙度要求是针对已经加工完的工件表面。

According to the standard, the marking and reading direction of surface roughness is consistent with that of size, as shown in Fig. 4-5.

标准规定表面粗糙度的标注和读取方向与尺寸的标注和读取方向一致,如图 4-5 所示。

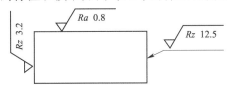

Fig. 4-5　Marking Direction of Surface Roughness Requirements
图 4-5　表面粗糙度要求的标注方向

The surface roughness requirements can be marked on the contour line, and the symbol should point from the outside of the material and contact the surface. If necessary, the surface roughness symbol can also be marked by a leader with an arrow or black dot, as shown in Fig. 4-6 and Fig. 4-7.

表面粗糙度要求可标注在轮廓线上,其符号应从材料外指向并接触表面。必要时,表面粗糙度符号也可用带有箭头或黑点的指引线引出标注,如图 4-6 和图 4-7 所示。

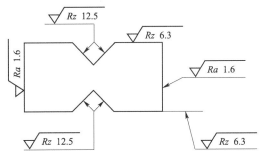

Fig. 4-6　Marking of Surface Roughness on Contour Line
图 4-6　表面粗糙度在轮廓线上的标注

117

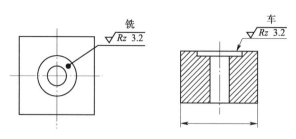

Fig. 4-7   Marking Surface Roughness with Leader
图 4-7   用指引线引出标注表面粗糙度

Surface roughness can be marked on a given size without causing misunderstanding, as shown in Fig. 4-8.

在不会引起误解时,表面粗糙度可以标注在给定的尺寸上,如图 4-8 所示。

The surface roughness requirements can be marked above the geometric tolerance frame, as shown in Fig. 4-9.

表面粗糙度要求可标注在几何公差框格的上方,如图 4-9 所示。

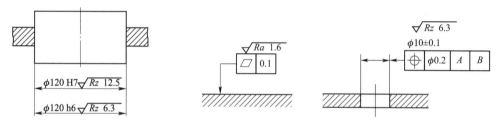

Fig. 4-8   Marking of Surface Roughness on size Line
图 4-8   表面粗糙度标注在尺寸线上

Fig. 4-9   Marking of Surface Roughness above Geometric Tolerance Frame
图 4-9   表面粗糙度标注在几何公差框格的上方

The surface roughness requirements can be directly marked on the extension line of the size line, or can be led out by the leader with arrow. The surface roughness of cylinder and prism is required to be marked only once, as shown in Fig. 4-10. For each prism surface with different surface roughness requirements, they should be marked separately, as shown in Fig. 4-11.

表面粗糙度要求可以直接标注在尺寸线的延长线上,或用带箭头的指引线引出标注。对于圆柱和棱柱表面的粗糙度要求只标注一次,如图 4-10 所示。对于每个棱柱表面用不同表面粗糙度要求的情况,则应分别单独标注,如图 4-11 所示。

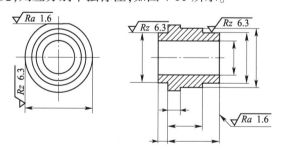

Fig. 4-10   Marking of Surface Roughness on Cylinder Feature Extension Line
图 4-10   表面粗糙度标注在圆柱特征的延长线上

Chapter 4　Surface Roughness 表面粗糙度

A simplified method for surface roughness. If most (all) surfaces of the workpiece have the same surface roughness requirements, the surface roughness requirements can be uniformly marked near the marking column of the drawing, as shown in Fig. 4-12 a) and b). The surface with the same surface roughness requirements can be simplified in the form of equation near the figure or title block with complete symbols with letters, as shown in Fig. 4-13.

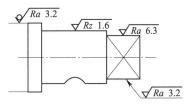

Fig. 4-11　Marking of Surface Roughness on Cylinder and Prism

图 4-11　圆柱和棱柱表面粗糙度的注法

表面粗糙度的简化注法。如果工件多数(全部)表面有相同的表面粗糙度要求,则其表面粗糙度要求可统一标注在图样的标注栏附近,如图 4-12a)和 b)所示。可用带字母的完整符号,在图形或标题栏附近,以等式的形式对有相同表面粗糙度要求的表面进行简化注法,如图 4-13 所示。

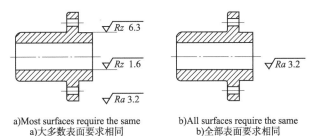

a) Most surfaces require the same
a) 大多数表面要求相同

b) All surfaces require the same
b) 全部表面要求相同

Fig. 4-12　Marking of Most or All Surfaces with the Same Surface Roughness Requirements

图 4-12　大多数或全部表面有相同的表面粗糙度要求的标注

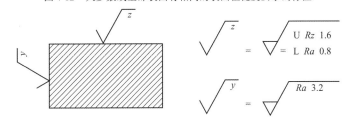

Fig. 4-13　Simplified Annotation for Limited Paper Space

图 4-13　在图纸空间有限时的简化标注

4.5　Measurement of Surface Roughness 表面粗糙度的检测

For surface roughness, if the direction of measurement section is not specified, the measurement should be carried out in the direction of maximum amplitude parameter, generally speaking, in the direction perpendicular to the surface processing texture. For the structure and operation method of the instrument used for measuring surface roughness, please refer to the experimental instruction. The measuring methods of surface roughness include the comparison method, the light cutting method, the interference method, the needle tracing method, the impression method and the three-dimensional geometric surface measurement method.

对于表面粗糙度,如未指定测量截面的方向,则应在幅度参数最大值的方向上进行测量,一般来说也就是在垂直于表面加工纹理方向上测量。测量表面粗糙度所用仪器的结构和操作方法可以参阅实验指导书。表面粗糙度的检测方法主要有比较法、光切法、干涉法、针描法、印模法和三维几何表面测量法等。

The comparison method is a method to determine the surface roughness by comparing the measured surface with the roughness template of known height parameters. The comparison method is simple and suitable for use in workshop. The accuracy of the judgment depends largely on the experience of the inspectors.

比较法是用被测表面与已知高度参数值的粗糙度样板相比较来确定表面粗糙度的一种方法。比较法较为简单,适合在车间使用。其判断的准确性在很大程度上取决于检验人员的经验。

The light cutting method is a method to measure surface roughness by using light cutting principle. The roughness value can be determined by measuring and data processing according to the definition of evaluation parameters. The optical section microscope is suitable for measuring Rz in the range of 0.8 ~ 80 μm.

光切法是利用光切原理来测量表面粗糙度的一种方法。按评定参数的定义进行测量和数据处理即可确定粗糙度的数值。光切显微镜适于测量 Rz,其测量范围为 0.8~80μm。

The interference method is to measure the surface roughness based on the principle of light wave interference. The common instrument is interference microscope. This instrument is suitable for evaluating the surface roughness with Rz value, and the measuring range is 0.025 ~ 0.8μm.

干涉法是用光波干涉原理来测量表面粗糙度。常用的仪器是干涉显微镜。这种仪器适宜于用 Rz 值来评定表面粗糙度,其测量范围为 0.025~0.8μm。

The needle tracing method uses the stylus to move directly on the measured surface, so as to measure the value of surface roughness. The electric profilometer is an instrument which uses the needle tracing method to measure the surface roughness. Generally, the Ra value is directly displayed, and the measurement range is 0.02 ~ 5μm.

针描法是利用触针直接在被测表面上移动,从而测出表面粗糙度的参数值。电动轮廓仪就是利用针描法测量表面粗糙度的仪器。通常直接显示 Ra 值,测量范围为 0.02~5μm。

The impression method can be used to evaluate the surface roughness of some parts, such as deep holes, blind holes, grooves and the inner surface of large parts, which are inconvenient to be measured directly by surface roughness instrument. The impression method refers to the method of printing the surface to be measured with plastic material, and then measuring the impression surface. Because the impression material can not completely fill the valley bottom and the shrinkage effect of the impression material, there is a certain difference between the measured roughness of the impression and the actual surface roughness of the part. Therefore, the measurement results should be corrected according to the experiment.

对于一些不便用表面粗糙度测量仪器直接测量的零件表面,如深孔、盲孔、凹槽以及大型零件的内表面等,可用印模法来评定其表面粗糙度。印模法是指用塑性材料将被测表面印下来,然后对印模表面进行测量的方法。由于印模材料不能完全填满谷底,且印模材料有收缩效应,所以测得印模的粗糙度与零件实际表面的粗糙度之间有一定差别。因此,一般应根据实验对测量结果进行修正。

The one-dimensional and two-dimensional measurement of surface roughness can only reflect some geometric characteristics of the surface roughness. It is not sufficient to use it as a statistical feature of the whole surface. Only by using three-dimensional evaluation parameters can the actual characteristics of the measured surface be truly reflected. Therefore, different countries are committed to the research and development of the three-dimensional geometric surface measurement technology, the optical fiber method, the microwave method and the electron microscope have been successfully applied to the measurement of three-dimensional geometric surface.

表面粗糙度的一维和二维测量,只能反映表面不平度的某些几何特征,把它作为表征整个表面的统计特征是很不充分的,只有用三维评定参数才能真实地反映被测表面的实际特征。为此,各国都在致力于研究开发三维几何表面测量技术,现已将光纤法、微波法和电子显微镜等测量方法成功地应用于三维几何表面的测量。

Exercises 4 习题 4

4-1　What are the effects of surface roughness on the working performance of parts?
4-1　表面粗糙度对零件的工作性能有何影响?
4-2　What is the significance of specifying the evaluation length?
4-2　规定评定长度有何意义?
4-3　What are the meanings of surface roughness evaluation parameters Ra and Rz?
4-3　表面粗糙度评定参数 Ra 和 Rz 的含义是什么?
4-4　What factors should be considered when selecting surface roughness parameter values?
4-4　选择表面粗糙度参数值时,应考虑哪些因素?
4-5　What are the common surface roughness detection methods?
4-5　常用的表面粗糙度检测方法有哪几种?

Chapter 5　Fundation of Geometrical Quantity Measurement 几何量测量基础

The mechanical manufacturing industry can not develop without measurement technology, and the development of measurement technology has promoted the development of modern manufacturing technology. In "designing, manufacturing, testing", measurement plays an extremely important role.

机械制造业的发展离不开测量技术,测量技术的发展推动了现代制造技术的发展。在"设计、制造、测试"这3个环节中,测量起着极其重要的作用。

5.1　Introduction 概述

Because machining error of components and parts is inevitable, advanced tolerance standards should be adopted to specify reasonable tolerances for geometric quantities of mechanical components and parts for the realization of components and parts interchangeability. Detection is the general term for measurement and inspection. But without the use of appropriate inspection measures, specified tolerances will exist in name only, and can not play a role, so the measurement technology and its development is inseparable from the development of manufacturing. Measurement technique during mechanical manufacturing, mainly referring to study the problems of measuring the geometrical quantities, such as the size, the form, the orientation, or the position of a feature in a workpiece, is the technological guarantee carrying out the quality standards. Whether a part is geometrically qualified or not, needs to be determined by measurement or inspection order.

由于零件加工误差是不可避免的,所以应采用严格的公差标准来规定机械零件的几何量公差,以实现零件的互换性。检测是测量和检验的总称,检测技术的发展与制造业的发展是密不可分的,如果没有适当的检测措施,规定的公差将形同虚设,不能发挥作用。机械制造中的测量技术,主要是研究测量工件上某一特征的尺寸、形状、方向或位置等几何量的问题,是实施质量标准的技术保障。零件几何形状合格与否,需要按照测量或检查顺序确定。

5.1.1　Concept of Measurement and Inspection 测量和检验的概念

Measurement is a process of experimentally obtaining one or more quantity values that can reasonably be attributed to a quantity. In other words, it is the process to obtain the multiple or fraction by comparing the measurand geometrical quantities (such as length, angle) to the standard quantity with unit. Let the measured geometry is L, the unit of measurement used is E, then their ratio is q.

Chapter 5 Fundation of Geometrical Quantity Measurement 几何量测量基础

测量是通过实验获得一个或多个可以合理归属于一个量的量值的过程。换句话说,就是将被测量的几何量(如长度、角度等)与带有单位的标准量进行比较,从而得到对比比值结果的过程。假设被测几何是 L,标准量是 E,那么他们的比值是 q。

$$q = L/E \tag{5-1}$$

The magnitude of the measured geometry L is equal to the magnitude of the resulting q multiplied unit E, namely

被测几何量 L 的大小等于得到的比值 q 乘以标准量 E 的大小,即

$$L = qE \tag{5-2}$$

Inspection is a process to judge whether a workpiece is qualified by indentifying whether the measurand is in a specified range. Usually it is not necessarily for measured specific value. Generally, constant value measuring tools are used.

检验是通过识别被测变量是否在规定的范围内,来判断工件是否满足要求的过程。通常不需要测量具体的值,一般用定值量具。

5.1.2 Measurement Processing 测量处理过程

Any geometric quantities are made up of two parts, numerical representing geometrical quantities and units of measurement of the geometrical quantities, such as 3.69m or 3.690mm. Obviously, any measurement, the measurand must be first clear and the unit of measurement should be determined, measurement methods must be followed to adapt the measurand, and the measurement results should achieve the required accuracy. That is, a whole geometric measurement process should include four features, i. e. measurand, unit, measurement method, and measurement precision.

任何几何量都由数值和单位两部分组成,数值代表几何量大小,单位为几何量的计量单位,如 3.69m 或 3.690mm。任何一项测量,首先必须明确测量对象与测量单位,选用适应测量对象的测量方法,并且测量结果应达到要求的精度。也就是说,一个完整的几何测量过程应包括被测量、单位、测量方法和测量精度 4 个特征。

Measurand refers to a geometrical quantity, i. e. length (include angle), surface roughness, geometrical errors, and the parameters of a screw thread or a gear and so on. In geometrical measurement, the unit of length include meter (m), millimeter (mm), micrometer(μm), the unit of angle include angle (°), minute ('), second (″) and so on.

被测量是指某一几何量,即长度(包括角度)、表面粗糙度、几何误差以及螺纹或齿轮的参数等。在几何测量中,长度的单位有米(m)、毫米(mm)、微米(μm),角度的单位有度(°)、分(')、秒(″)等。

Measurement method is a general term of measurement principle, measurement tools and measurement conditions. The measurement conditions refer to the environment during measuring, such as temperature, humidity, vibration, dusts, etc. Usually the measurement tools, reasonable methods is determined according to the requirements of precision, amplitude, heavy, amount etc.

测量方法是测量原理、测量工具和测量条件的总称。测量条件是指测量过程中所处的环境,如温度、湿度、振动、灰尘等。测量工具、方法则根据精度、幅度、重量、数量等要求来

确定。

The measurement precision refers to the closeness of the measurement result to the true value, the corresponding concept is measurement error. The measurement error is inevitable as there are lots of factors affecting the measurement result. The larger the measurement error is, the lower the measurement precision is, and the lower the measurement error is, the higher the measurement precision is. Measurement without precision is meaningless.

测量精度是指测量结果与真值的接近程度,对应的概念是测量误差。由于影响测量结果的因素很多,测量误差不可避免。测量误差越大,测量精度越低,反之,测量误差越小,测量精度越高,没有精度要求的测量是毫无意义的。

5.2　Measurement Standard, Datum and Data 测量标准、基准和数据

5.2.1　Standard of Length and Angle 长度和角度的标准

Measurement needs standards in production and scientific experiment, and the amount of measurement quantity indicated by the standard need to be provided by the datum standard, in order to ensure the precision of measurement. Therefore, it is necessary to establish a unified, reliable basic unit of the measurement. In the field of geometrical measurement, measuring datum can be divided into length and angle of the datum standard.

在生产和科学实验中,测量需要标准,以测量基准作为标准,可以确保测量的精确。因此,有必要建立一个统一、可靠的测量基本单元。在几何领域的测量中,测量基准可分为长度和角度的标准。

As mentioned above, in order to ensure the precision of measurement of length, first of all a unified, reliable length standard needs to be established during the measurement. In the international unit of systems, the basic unit of length is meter (m). In China, the international unit system is adopted. In the mechanical engineering, the commonly used units include millimeter (mm), micrometer (μm) and nanometer (nm). Relationships among them are: 1m = 1000mm; 1mm = 1000μm; 1μm = 1000nm. In the engineering drawings, the default unit for size and geometrical tolerance is mm, and the unit of the amplitude of surface roughness is μm. Precision measurement unit is commonly μm, ultra-precision measurement unit is commonly nm.

为了保证长度测量的准确性,首先在测量时需要建立一个统一、可靠的长度标准。在国际单位制中,长度的基本单位是米(m),中国采用国际单位制。在机械工程中,常用的单位有毫米(mm)、微米(μm)和纳米(nm)。它们之间的关系为:1m = 1000mm;1mm = 1000μm;1μm = 1000nm。在工程图纸中,尺寸和几何公差默认单位为mm,表面粗糙度幅值单位为μm。精密测量单位一般用μm,超精密测量单位一般用nm。

Angle is one of important geometric quantities. The central angle of a circle is defined as 360°, therefore it is necessary to establish a natural datum like the standard of length. Common angle measurement unit is radian (rad), micro arc (μrad) and degrees, minutes and seconds.

Chapter 5 Fundation of Geometrical Quantity Measurement 几何量测量基础

角度是重要的几何量之一。圆的圆心角定义为360°,需要建立一个像长度单位的参考标准。角度的测量单位为弧度(rad)、微弧(μrad)、度数、分、秒。

5.2.2 Transmission System of Length and Angle 长度和角度的传递

In practice, light can not be used directly as the length datum for measuring the length, therefore accurate dissemination system need to be established, then a variety of measuring instruments are used to measure. In order to keep measurement traceability, the magnitude of the length of the datum must be accurately transmitted to the measurement instruments used in the production and the measured workpiece. Length measurement is divided into two parallel systems to pass down, one is the end measurement (gauge block), and the other is the line gauge (linear scale).

在实践中,光不能直接作为测量长度的基准,因此需要建立精确的传递系统,然后使用各种测量仪器进行测量。为了保持测量的可追溯性,必须将基准长度的大小准确地传递给生产中使用的测量仪器和被测工件。长度测量分为两个平行系统向下传递,一个是端部测量(量规块),另一个是线规(线性标尺)。

5.2.3 Gauge Blocks and Their Applications 量块及其应用

Gauge blocks are standard devices often used in precise measurement, they are divided into length and angle gauge blocks. Here we talk about the length gauge blocks.

量块是精密测量中常用的标准装置,分为长度量块和角度量块。这里讲的是长度量块。

A gauge block is a block of metal or ceramic with two opposing faces ground precisely flat and parallel, a precise distance apart. Standard gauge blocks are made of a hardened steel alloy, while calibration gauge blocks are often made of tungsten carbide or chromium carbide because it is harder and wears less. Gauge blocks come in sets of blocks of various lengths, along with two wear blocks, to allow a wide variety of standard lengths to be made up by stacking them. The length of each block is actually slightly shorter than the nominal length stamped on it, because the stamped length includes the length of one wring film, a film of lubricant which separates adjacent block faces in normal use.

量块是一块金属或陶瓷块,两相对表面被精确地打磨成平面或平行平面,彼此之间有精确的相对位置。标准量块由淬硬合金钢制成,而校准量块通常由碳化钨或碳化铬制成,硬度高,磨损少。块规是一组不同长度的量块,通过堆叠它们来产生各种标准长度。每个量块的长度实际上略短于量块上所印制的公称长度数值,因为印制的长度值包括一个油膜的长度。在正常使用中,油膜是一种将相邻量块表面分开的润滑剂。

In use, the blocks are removed from the set, cleaned of their protective coating (petroleum jelly or oil) and wrung together to form a stack of the required size, with the minimum number of blocks. Gauge blocks are calibrated to be accurate at 20 °C and should be kept at this temperature when taking measurements. This mitigates the effects of thermal expansion. The wear blocks made of a harder substance like tungsten carbide, are included at each end of the stack, whenever possible, to protect the gauge blocks from being damaged in use.

在使用中,取出量块清洁其保护涂层(凡士林或油),形成一个所需尺寸的堆叠。量块经过校准,在 20°C 下保持准确,测量时应保持这个温度,以减小热膨胀的影响。磨损块由硬质合金等硬质物质制成,尽可能地将其放在堆叠的两端,起到保护量规块的作用。

If the size of gauge block is less than 10 mm, the section size is 30 mm × 9 mm. If size is 10 ~ 1000 mm, the section size is 35 mm × 9 mm.

如量块尺寸小于 10mm,截面尺寸为 30mm×9mm。若尺寸在 10 ~ 1000mm,断面尺寸为 35mm×9mm。

In order to meet the needs of different occasions, gauge blocks can be made into different accuracy grades. According to the manufacturing precision and calibration accuracy, the accuracy of the gauge blocks is defined as a number of "grade" and "level".

为了满足不同场合的需要,量块可制成不同的精度等级。根据量块的制造精度和校准精度,将量块的精度定义为若干"等级"和"水平"。

Gauge block "grade" and "level" are two different forms of the accuracy at the view of mass manufacture and individual calibration. In accordance with the "grade" the nominal size marked on the gauge block is used as the working size, which includes its manufacturing error. According to "level", the actual size after verification is used as the working size, the size does not contain the manufacture error, but includes the measurement error of the calibration. For the same gauge block, calibration measurement error is much smaller than manufacturing errors. Therefore, the accuracy when used according to "level" higher than that when used according to "grade", and the service life can be prolong while keeping the original precision of the gauge block.

量块的"等级"和"水平"是批量生产和个别校准对的精度。根据"等级",用量块上标注的公称尺寸作为工作尺寸,其中包括其制造误差。根据"水平",检定后的实际尺寸作为工作尺寸,该尺寸不包含制造误差,但包含校准后的测量误差。对于同一量块,标定测量误差比制造误差小得多。因此,按"水平"使用精度高于按"等级"使用精度,并可在保持量块原有精度的基础上延长使用寿命。

5.3 Metrological Equipments and Measurement Methods
计量设备和测量方法

5.3.1 Metrological Equipments 计量设备

Metrological equipments are measurement tools, measurement instruments, and other general terms used for the purpose of measuring technology devices.

计量设备是指测量工具、测量仪器以及其他用于测量技术目的的通用设备。

Measurement tools refer to the metrological equipment that reproduce the magnitude in the fixed form. Measurement tool features are: generally there is no indicator, and measurement tools do not include the moving measurement feature in the measurement process. A tool for reproducing single value called the single value measurement tool, such as gauge blocks, square and so on. A tool can be used to replicate a range of different values within a certain range, called general measurement tool. General measurement tools according to their structural features are divided into

Chapter 5 Fundation of Geometrical Quantity Measurement 几何量测量基础

the following categories: fixed line gauge, such as ruler, tape measure, etc. Vernier gauge, such as vernier caliper, vernier depth scale, vernier angular scale. Screw micrometer gauge, such as the inner, outside micrometer and spiral micrometer. There are special measurement tools without line to inspect the size error or geometrical errors, which only used to judge the workpieces pass or fail and geometrical quantities can not be achieved, such as plain limit gauges, positional gauges, screw gauges.

量具是指以固定形式再现量值的计量设备,如单值测量工具、量块、方尺等。通用测量工具有直尺、卷尺等。游标量具有游标卡尺、游标深度尺、游标角度尺等。螺杆千分尺有内千分尺、外千分尺和螺旋千分尺。还有一些量具可以用来检测工件尺寸误差或几何误差,判断工件的合格与否,但是无法获得几何量测量值,如普通极限量规、定位量规、螺纹量规等。

Instruments refer to the measurement instruments that the measured geometric quantity can be converted into directly observable indication or equivalent information. The instrument itself includes a movable measuring feature and can be indicative of a specific measured value. According to the principle of the original signal conversion measurement instruments can be divided into the following types.

仪器是指被测几何量能够转化为直接示值或等效示值的测量仪器。仪器本身包括一个可移动的测量元件,用来确定具体的测量值。根据信号转换的原理可将测量仪器分为以下几种。

Mechanical instruments refer to the instruments that the mechanical method is used to implement the original signal, generally with a mechanical micrometric mechanism. These instruments have the advantages of simple structure, stable performance, convenient use, such as the indicator, the lever comparator and so on.

机械仪器是指用机械方法测量原始信号的仪器,一般带有机械测微机构。这些仪器具有结构简单、性能稳定、使用方便等优点,包括指示器、比较仪等。

Optical instruments refer to the instruments that the optical method is used to implement the original signal, generally with optical amplification (micrometer) mechanism. These instruments have the advantages of high precision, stable performance, such as the optical comparator, the tool microscope, the interferometer and so on.

光学仪器是指用光学方法来测量原始信号的仪器,一般带有光放大(测微计)机构。这些仪器具有精度高、性能稳定的优点,包括光学比较仪、工具显微镜、干涉仪等。

Electrical instruments refer to the instruments that the original signal can be converted to electric signal, usually with amplification, filter circuit, etc. These instruments with high precision, measurement signal through the analog/digital (A/D) conversion, easy to interface with the computer, realize the automation of measurement and data processing, such as inductance comparator, electric profilometer, roundness instrument, and so on.

电动仪表指的是能将原始信号转换成电信号的仪表,通常带有放大、滤波电路等。这些仪器测量精度高,测量信号通过模拟/数字(A/D)转换,易于与计算机接口,实现测量和数据处理的自动化,包括电感比较仪、电轮廓仪、圆度仪等。

Pneumatic instruments are the measuring instruments using the compressed air as medium, change through the pneumatic system flow or pressure to achieve the original signal conversion. This instrument has the advantages of simple structure, high measuring accuracy and efficiency, convenient operation, but the indication range is small, such as water column type pneumatic measuring instrument, float type pneumatic measuring instrument, etc.

气动仪表是用压缩空气作为介质,通过改变气动系统的流量或压力来实现原有信号转换的测量仪表。仪器的特点是结构简单,测量精度和效率高,操作方便,但指示范围小,包括水柱式气动测量仪、浮子式气动测量仪等。

The metering device is defined as the overall of necessary measurement apparatus and auxiliary equipment to determine the measurand geometric quantity value. It is capable of measuring the same workpiece with several geometric features and a complex shape, contributing to the detection automated or semi-automated, such as comprehensive precision gear tester, engine cylinder hole geometry precision measuring instrument, etc. The basic technical performance index of metrological equipment is an important basis for its rational selection and use.

计量装置是指由必要的测量仪器和辅助设备来确定被测几何量值。它能够测量同一工件的多个特征和复杂形状的工件,有助于半自动化或自动化的检测。如精密齿轮综合测试仪、发动机气缸孔几何精密测量仪等。计量装置的基本技术性能指标是计量装置合理选择和使用的重要依据。

5.3.2　Measurement Methods 测量方法

Measurement methods can be classified in a different form. Absolute measurement (direct comparison measurement) is that measurement indication can directly indicate the full value of the measured size. For example, we use vernier calipers, micrometers, length measuring instrument and so on to measure the diameter of the shaft. Relative measurement (comparative measurement, differential measurement) is that measurement indication represents only the deviation or ratio of the measured size to the known standard value. In Layman's terms, relative measurement is measuring something compared to another thing, or estimating things proportionally to one another. Measured results are equal to the algebraic sum or the product of the value of measured size and the known standard value. Generally speaking, the relative measurement method with high precision and simple operation, it is widely used in precision length measurement.

测量方法可以分为绝对测量和相对测量。绝对测量(直接测量)是指测量可以直接获得被测尺寸的全部值。例如,我们用游标卡尺、千分尺、测长仪等来测量轴的直径。相对测量(比较测量)是指获得被测尺寸与已知标准值的偏差或比值测量的。相对测量是被测对象与参考基准的比较,或按比例估计被测对象,被测结果等于被测尺寸值与已知标准值的代数和或者乘积。一般来说,相对测量方法具有精度高、操作简单等优点,被广泛应用于精密长度测量中。

5.3.2.1　Direct and Indirect Measurement 直接测量和间接测量

According to whether the real measured value is measurand, the measurement methods can be divided into direct and indirect measurement.

Chapter 5　Fundation of Geometrical Quantity Measurement 几何量测量基础

根据实际被测值是否被直接测量,测量方法可分为直接测量法和间接测量法。

Direct measurement refers to measuring exactly the thing that you're looking to measure. There's no function relationship between the quantity to be measured and other real measured quantity, and the quantity to be measured can be obtained directly. The indication we got on the measurement instruments can be the full value of the measured size, and can also be standard deviations. For example, we measure length using line scale or gauge block. Generally, direct measurement is simple, without complex functions calculation.

直接测量指的是精确地测量想要测量的几何量的数值。被测量与其他实际被测量之间不存在函数关系,可以直接得到被测量。我们在测量仪器上得到的示值可以是测量尺寸的全值,也可以是标准差。例如,我们用线尺或量块测量长度。一般情况下,直接测量简单,没有复杂的函数计算。

Indirect measurement means that we're measuring something by measuring something else. There is function relationship between the quantity to be measured and other real measured quantity, and the quantity to be measured should be obtained after function aided calculation. Generally speaking, indirect measurements are more troublesome. When the measured size can't be measured directly or the direct measurement does not reach the required precision, we often have to use indirect measurement. In practice, measuring an angle, a taper, a hole pitch, the curvature radius of an arc and the relevant size of intersection commonly use indirect measurement method.

间接测量指的是通过测量某个值来得到实际要测的值。待测量与其他实际被测量之间存在函数关系,须经过函数辅助计算后得到待测量。一般来说,间接测量比较麻烦。当被测尺寸不能直接测量或直接测量不能达到所要求的精度时,我们常常不得不采用间接测量。在实际测量中,测量一个角度、一个锥度、一个孔距、一个圆弧的曲率半径及其相关交点的位置,一般采用间接测量方法。

5.3.2.2　Single Parameter Measurement or Comprehensive Measurement 单参数测量或综合测量

According to whether there are multiple geometrical quantities of the workpiece to be measured at the same time, measurement methods can be divided into single parameter measurement and comprehensive measurement.

根据被测工件是否同时测量多个几何量,将测量方法分为单参数测量和综合测量。

Single parameter measurement refers to the various parameters to be measured separately and without contact with each other. For example, we respectively measure the pitch diameter, pitch and flank angle of thread and so on. In General, single parameter measurement is less efficient. For high precision parts or for process analysis, single parameter measurement should be adopted.

单参数测量指单独测量各个互不相关的参数。例如,我们分别测量螺纹的节径、节距和侧倾角等。一般情况下,单参数测量效率较低。对于精度较高的零件或过程分析,应采用单参数测量。

Comprehensive measurement is that by measuring the comprehensive parameter associated with several parameters of the parts, we comprehensively judge whether the part is qualified. For example we use a thread gauge to examine the threaded. Comprehensive measurement method has the advantage of high efficiency, suitable for mass production.

综合测量通过测量与零件若干参数相关联的综合参数,综合判断零件是否合格。比如我们用螺纹量规来检查螺纹。综合测量方法具有效率高,适合大批量生产的优点。

5.3.2.3 Contact or Non-contact Measurement 接触和非接触测量

According to whether there is mechanical forces between the object to be measure and the probe of measurement instrument, the measurement methods can be divided into contact and non-contact measurement.

根据被测物体与测量仪器探头之间是否存在机械力,将测量方法分为接触式测量和非接触式测量。

Contact measurement means that the sensitive components of measurement instrument are in direct contact with surface of the parts to be measured. For instance, we use vernier caliper to measure diameter. The characteristics of the contact measurement method is the existence of measuring force, it can make reliable contacts. On the contrary, it will also cause the deformation of the measurement instruments and the measured parts, resulting in measurement error.

接触测量指测量仪器的敏感元件与被测零件表面直接接触的测量方法。例如,用游标卡尺测量直径。接触测量法的特点是存在测量力,它能使接触可靠。同时,还会造成测量仪器和被测零件的变形,造成测量误差。

Non-contact measurement refers to the method that the sensitive components of measurement instrument are not in direct contact with the surface of the parts to be measured. For instance, measurement of parts by using the projection method and optical interference.

非接触式测量指测量仪器的敏感元件与被测零件表面不直接接触的方法。例如,用投影法和光干涉法测量零件。

5.3.2.4 Passive and Active Measurement 被动测量和主动测量

According to whether the measurement is during the process of machining, the measurement method can be divided into passive or active measurement.

根据测量是否在加工过程中进行,测量方法可分为被动测量和主动测量。

Passive measurement (offline measurement) refers to the process of measuring parts after finishing machining. Its role is only to discover and identify waste.

被动测量(离线测量)是指工件加工后的测量过程。它的作用是发现和识别废品。

Active measurement (online measurement) means that according to the measurement results the process of measuring parts during machining, directly control the machining process in real time to decide whether to continue processing/machining or need to adjust the machine. Its role is only to discover and identify waste. So it can prevent the occurrence of waste products. Because it is measured in the process of machining, it can also shorten the parts production cycle.

主动测量(在线测量)是指在加工过程中对零件进行测量的过程。主动测量直接实时控制加工过程,以决定是否继续加工或需要对机床进行调整后加工。它的作用是在加工过程中发现和识别误差,防止产生废品。在线测量是在加工过程中进行的,可以缩短零件的生产周期。

Chapter 5   Fundation of Geometrical Quantity Measurement 几何量测量基础

There are many classifications of measurement methods. For example, according to whether the measuring conditions are changed in the process, the measurement methods can be divided into equal precision and non-equal precision measurement. Here is no longer the detail. The above classifications of the measurement are considered from different perspectives, however, for a measurement of a specific process, it may have characteristics of several methods of measurement. Using three-coordinate measuring machine to measure the contour of a workpiece, includes direct, contact, active measurement and so on. Therefore, the choice of measurement method should consider the structural features of the object being measured, the precision requirement, production batch, technical conditions and economic benefits, etc.

测量方法有许多分类。例如,根据测量条件在过程中是否发生变化,测量方法可分为等精度测量和非等精度测量,这里不再赘述。以上对测量的分类是从不同的角度来考虑的,但是对于一个特定过程的测量,它可能具有几种测量方法的特点。利用三坐标测量机测量工件轮廓,包括直接测量、接触测量、主动测量等。因此,测量方法的选择应考虑被测对象的结构特点、精度要求、生产批次、技术条件和经济效益等因素。

5.4 Measurement Error and Data Processing
测量误差及数据处理

The purpose of measurement is to provide information about a quantity of interest measurand. For example, the measurand might be the volume of a vessel, the potential difference between the terminals of a battery, or the mass concentration of lead in a flask of water.

测量的目的是获取被测对象的准确值。例如,测量可以是容器的体积,电池两端之间的电位差,或者是烧瓶中铅的质量浓度。

No measurement is exact. When a quantity is measured, the outcome depends on the measuring system, the measurement procedure, the skill of the operator, the environment, and other effects. Even if the quantity were to be measured several times, in the same way and in the same circumstances, a different measured value would in general be obtained each time, assuming that the measuring system has sufficient resolution to distinguish between the values.

没有绝对精确的测量。当一个几何量被测量时,结果取决于测量系统、测量程序、操作者的技能、环境和其他因素的影响。假设测量系统有足够的分辨率,即使用同样的方法和在同样的情况下,对同一几何量测量多次,每次都会得到不同的测量值。

The dispersion of the measured values would be related to how well the measurement is made. Their average would provide an estimate of the true value of the quantity that generally would be more reliable than an individual measured value. The dispersion and the number of measured values would provide information relating to the average value as an estimate of the true value. However, this information would not generally be adequate.

测量值的离散度决定测量结果的可靠性。多次测量的平均值提供了一个真实值的估计值,通常比单个测量值更可靠。作为对真实值的估计,平均值取决于测量值的离散度和测量样本数量。但是,通常情况下,样本的数量是不够充足的。

Measurement error (also called as observational error, or error of measurement, shortly error) can be defined as measured quantity value minus a datum quantity value. The absolute error is the difference between the measured value (the approximation) and the true (exact) value with the same units. The relative error is the ratio of the absolute value of the absolute error to the true value, the relative error is a dimensionless number, usually expressed as a percentage.

测量误差(也称观测误差,简称误差)定义为被测量值减去参考量值。绝对误差是测量值(近似值)和真实值(精确值)之间的差值,测量单位保持一致。相对误差是绝对误差的绝对值与真实值的比值,是无量纲数,通常用百分数表示。

Due to the presence of measurement error, the measured values can only approximately reflect the truth values of geometrical quantities to be measured. In order to reduce measurement error, it must be to analyze the cause of measurement error to improve the measurement accuracy. In actual measurements, there are many factors that cause measurement errors, summed up in the following aspects.

由于测量误差的存在,测量值只能近似反映被测几何量的真实值。为了减少测量误差,必须分析测量误差产生的原因,提高测量精度。在实际测量中,造成测量误差的因素有很多,主要有以下几个方面。

The errors of measurement instruments refer to the errors of the measurement instruments themselves, including the various errors during design, manufacture and use of measurement instruments, and the sum of the errors reflects on errors of indication and repeatability of measurement.

测量仪器的误差指测量仪器本身的误差,包括测量仪器在设计、制造和使用过程中的各种误差,这些误差的总和反映在测量的示值误差和测量的重复性上。

The errors of measurement method state that imperfect measurement methods (including the calculation formula is inaccurate, inappropriate choice of measurement method, the workpiece mounted position is not accurate, etc.) can resulting in the errors. For example, during the contact measurement, the measurement device and the parts to be measured is deformed because the impact of probe measurement force to generate a measurement error.

测量方法误差指测量方法不完善(包括计算公式不准确、测量方法选择不恰当、工件安装位置不准确等)产生的误差。例如,在接触测量过程中,测量装置和被测零件由于探头测量力的影响而产生变形,产生测量误差。

The environmental error refers to the error caused by the environmental conditions that do not meet the standards when measuring. Personnel errors are man-made errors by measuring staff, such as measuring targeting inaccurate readings or estimated reading errors, etc., will have artificial measurement error. According to the characteristics and properties of the measurement errors, they can be divided into three kinds: systematic error, random error, and gross error.

环境误差是指测量时由于环境条件不符合标准而引起的误差。人员误差是由测量人员人为造成的误差,如测量目标读数不准确或估计读数误差等,都会产生人为的测量误差。根据测量误差的特点和性质,可将测量误差分为系统误差、随机误差和粗大误差3种。

Systematic error means that as under certain measurement conditions, for the same measurant in replicate measurements, the symbol and magnitude of errors remain constant, or errors varies in a predictable manner. The former is called definite systematic errors. For example, when we tailor-made the comparator using gauge blocks according to the nominal size, their manufacturing company errors would produce the measurement errors; and micrometer zeroed incorrectly would resulting in measurement errors. The latter is called variable systematic error.

系统误差指在一定的测量条件下,对于重复测量的同一被测量对象,一直存在的误差。误差的符号和大小保持不变,称为定值系统误差。例如,量具的制造误差带来的测量误差就属于定制系统误差。千分尺调零不正确导致的测量误差称为变值系统误差。

Random error means that as under certain measurement conditions, for the same measurant in replicate measurements, the symbol and magnitude of errors varies in an unpredictable manner. Random errors are mainly due to some random or uncertain factors in the process of measuring. All the errors, for example, caused by clearance and friction of transmission mechanism, force of the unstable measurement, as well as the temperature fluctuation and so on, belong to the random errors.

随机误差是指在一定的测量条件下,对同一被测对象进行重复测量时,误差的符号和大小以不可预测的方式变化。随机误差主要是由于测量过程中的一些随机或不确定因素造成的。传动机构间隙和摩擦、不稳定的测量力以及温度波动等引起的误差都属于随机误差。

Gross error refers to the measure under certain conditions beyond the expected measurement error, that is, the measurement error of the measurement result is distorted. The measured value containing a gross error is called outliers, the error is relatively large. Gross error's generation has subjective reasons such as reading error caused by measuring staff negligence, and objective reasons such as measurement error caused by the vibration of an external sudden. Since the gross error obviously distort the measurement results, when dealing with the measurement data, we should eliminate it according to the discrimination criteria of gross error.

粗大误差是指在一定条件下的测量超出了预期的测量误差,即测量结果的误差被扭曲。含有粗大误差的测量值称为离群值,误差值比较大。粗大误差的产生既有主观原因又有客观原因,主观原因如测量人员的疏忽造成的读数误差,客观原因如外部突然振动造成的测量误差。由于粗大误差对测量结果的扭曲明显,因此在对测量数据进行处理时,应将其判别并进行剔除。

The measurement accuracy refers to the proximity of the measured value and its true value. It is the same concept as the measurement error from two different perspectives. The larger the measurement error is, the lower the measurement accuracy is; the smaller the measurement error is, the higher the measurement accuracy is. In order to reflect the different effects of system error and random error on the measurement results, the measurement accuracy can be divided into the following types:

测量精度是指测量值与真实值的接近程度,与测量误差是同一个概念。测量误差越大,测量精度越低;测量误差越小,测量精度越高。为了反映系统误差和随机误差对测量结果的不同影响,测量精度可分为以下几种:

Precision is the closeness of agreement between indications or measured quantity values obtained by replicate measurements on the same or similar objects under specified conditions. Measurement precision is inversely related to random measurement errors, but is not related to systematic measurement errors.

精密度是指在规定条件下,对相同或类似物体进行重复测量得到的指征或测量值之间的接近程度。精密度与随机测量误差呈负相关关系,与系统测量误差无关。

Trueness is the closeness of agreement between the average of an infinite number of replicate measured quantity values and a datum quantity value. Measurement trueness is inversely related to systematic measurement error, but is not related to random measurement error.

测量真实度是无数个重复测量的量值的平均值与参考量值的接近程度。测量真实度与系统测量误差呈负相关关系,与随机测量误差无关。

Percent of accuracy is the closeness of agreement between a measured quantity value and a true quantity value of a measurand. Measurement accuracy is inversely related to both systematic measurement error and random measurement error.

准确度是被测量的量值与被测量物的真实量值之间接近一致的关系。测量精度与系统测量误差和随机测量误差均呈负相关关系。

Exercises 5 习题 5

5-1 What is the essence of the measurement? What features are included in a measurement process?

5-1 测量的本质是什么?测量过程包括哪些特性?

5-2 What is the legal unit of length?

5-2 长度的法定单位是什么?

5-3 What is the difference between indication range and measurement range? Give an example.

5-3 示值范围和测量范围有什么区别?请举例说明。

5-4 Give some examples to explain the differences between absolute measurement and relative measurement, direct and indirect measurement.

5-4 举例说明绝对测量与相对测量、直接测量与间接测量的区别。

5-5 Please describe the types, properties of measurement errors.

5-5 描述测量误差的种类、性质。

Chapter 6　Tolerances and Fits for Typical Parts 典型零部件的公差与配合

6.1　Cylindrical Gear 圆柱齿轮

6.1.1　Application Requirements 使用要求

The gear transmission is widely used in machines and instruments. The gear transmission has the advantages that it has high transmission accuracy, wide range of transmission ratio, can realize arbitrary two shaft transmission in space such as intersecting shaft, and it has reliability and long service life, but also has the disadvantages of high environmental requirements and not too large wheelbase.

齿轮传动是机器和仪器中应用极为广泛的一种传动方式。齿轮传动具有传动精度高、传动比范围大、可以实现相交轴等空间任意两轴传动、工作可靠、使用寿命长等优点，也存在对环境要求高、轴距不能太大等缺点。

The quality and efficiency of gear transmission mainly depends on the manufacturing accuracy of gear and the installation accuracy of gear pair. The working performance, bearing capacity, service life and working accuracy are closely related to the manufacturing accuracy of gears. In order to ensure the accuracy and interchangeability of gear transmission, it is necessary to specify the tolerance of the gear and gear pair.

齿轮传动的质量和效率主要取决于齿轮的制造精度和齿轮副的安装精度。工作性能、承载能力、使用寿命和工作精度等都与齿轮的制造精度有密切的联系。为了保证齿轮传动的精度和互换性，需要规定齿轮的公差和齿轮副的公差。

As a transmission part, the application requirements of gear mainly include accuracy of transmission, stability of transmission, uniformity of load distribution and rationality of transmission backlash.

齿轮作为传动件，使用要求主要包括传动的准确性、传动的平稳性、载荷分布的均匀性和传动侧隙的合理性。

The accuracy of transmission requires the gear to be in a certain speed range, and the deviation of the maximum angle should be limited in a certain range, so as to ensure the coordinated movement of the driven and driving parts.

传动的准确性要求齿轮在一定转速范围内，最大转角的偏差应该限定在一定范围内，从而保证从动件与主动件的运动协调一致。

Transmission stability requires that the range of instantaneous transmission ratio of gear transmission is small. The change of the instantaneous transmission ratio which affects transmission stability is caused by gear single tooth error.

传动的平稳性要求齿轮传动的瞬时传动比变化范围较小,影响传动平稳性的瞬时传动比的变化是由齿轮单齿误差引起的。

The uniformity of load distribution requires good contact between tooth surfaces during gear meshing to avoid stress concentration and local wear of tooth surface, thus affecting the bearing capacity and service life of gears.

载荷分布的均匀性要求齿轮啮合时齿面接触良好,避免出现应力集中和造成齿面局部磨损,从而影响齿轮的承载能力和使用寿命。

The rationality of transmission backlash requires a certain clearance between the non working tooth surfaces when the gears are engaged. The gear pair should have proper side clearance, which is used to store lubricating oil, compensate for thermal deformation and elastic deformation, prevent tooth surface ablation or seizure during gear operation, and ensure normal operation of gear pair.

传动侧隙的合理性要求齿轮啮合时,非工作齿面间应具有一定的间隙。齿轮副应具有适当的侧隙,用来储存润滑油、补偿热变形和弹性变形,防止齿轮在工作中发生齿面烧蚀或卡死,保证齿轮副能够正常工作。

The application and specific working conditions of gear transmission are different, so the requirements of gear transmission are also different. The indexing gear of precision machine tool has the characteristics of small transmission power, small modulus and low speed, which requires high accuracy of gear transmission and low requirement of load distribution uniformity; if the gear is positive and reverse, the transmission backlash should be reduced as far as possible. The high-speed power gear on steam turbine has strict requirements for transmission accuracy, stability and load distribution uniformity, in addition to having enough large tooth side clearance. There are strict requirements for the working stability of automobile variable speed gears. The low-speed power gear of mining machinery requires that the tooth surface of meshing gear contacts well and the load distribution is uniform, and the requirements for transmission accuracy and stability are relatively low.

齿轮传动的用途和具体工作条件不同,对齿轮传动的使用要求也各有侧重。精密机床的分度齿轮具有传动功率小、模数小和转速低的特点,对齿轮传动的准确性要求高,对载荷分布均匀性的要求低;如果齿轮为正反转,还应尽量减小传动侧隙。汽轮机上的高速动力齿轮对传动的准确性、平稳性和载荷分布的均匀性都有严格的要求,此外还要有足够大的齿侧间隙。汽车变速齿轮对工作平稳性有严格要求。矿山机械的低速动力齿轮,要求啮合齿轮齿面接触良好、载荷分布均匀,对传动的准确性和平稳性要求相对低一些。

6.1.2　Error Factors　误差因素

There are four requirements for gear transmission: the accuracy of transmission, the stability of transmission, the uniformity of load distribution and the rationality of transmission backlash. The error that affects the requirement of gear transmission mainly comes from the machining error

Chapter 6　Tolerances and Fits for Typical Parts 典型零部件的公差与配合

of gear.

齿轮传动有4个使用要求：传动的准确性、传动的平稳性、载荷分布的均匀性和传动侧隙的合理性。影响齿轮传动使用要求的误差，主要来自齿轮的加工误差。

The main error that affects the transmission accuracy is the uneven gear pitch distribution that causes a cycle error of the gear rotation, mainly from the gear geometric eccentricity and motion eccentricity. Geometric eccentricity means that the shaft of the datum hole of the gear blank does not coincide with the shaft of the mandrel on the machine tool table. Motion eccentricity refers to the error that geometric eccentricity of indexing worm gear is reflected on the cutting gear, resulting in uneven distribution of pitch on the circle with gear datum shaft as the center. The accuracy of gear transmission should be evaluated by the maximum angle deviation caused by the uneven distribution of pitch caused by geometric eccentricity and motion eccentricity.

影响传动准确性的主要误差，是齿轮齿距分布不均匀而产生的周期为一转的周期误差。周期误差的主要原因是齿轮的几何偏心和运动偏心。几何偏心是指齿轮毛坯基准孔轴线与机床工作台芯轴轴线不重合。运动偏心是指机床分度蜗轮几何偏心复映到被切齿轮上的误差，造成与齿轮基准轴线同心的圆周上齿距分布不均匀。以几何偏心和运动偏心综合造成的各个齿距分布不均匀而产生的转角偏差的最大值，来评定齿轮传动准确性的精度。

The main error affecting the transmission stability is the error caused by the single tooth error of the gear, which takes a pitch angle as the period, mainly includes the single pitch deviation, that is, the pitch deviation between adjacent tooth profiles on the same side of the gear and the total deviation of the gear profile. The single pitch deviation is the algebraic difference between the actual pitch and the theoretical pitch on the end plane, which is close to a circle concentric with the gear shaft. The total deviation of tooth profile refers to the distance between two designed tooth profile traces containing the actual tooth profile trace within the range of tooth profile calculation value. When a gear rotates a tooth, the deviation of single pitch and the total deviation of tooth profile exist at the same time, so the accuracy evaluation of gear transmission stability needs comprehensive consideration.

影响传动平稳性的主要误差是齿轮的单齿误差引起的以一个齿距角为周期的误差，主要包括单个齿距偏差，即齿轮同侧相邻齿廓间的齿距偏差和齿轮齿廓的总偏差。单个齿距偏差是在端平面上，接近齿形中部的一个与齿轮轴线同心的圆周上，实际齿距与理论齿距的代数差。齿廓的总偏差是指在齿廓计算值范围内包容实际齿廓迹线的两条设计齿廓迹线间的距离。齿轮每转过一个齿，单个齿距偏差和齿廓总偏差是同时存在的，因此齿轮传动平稳性的精度评定需要综合考虑。

The main error affecting the uniformity of load distribution is tooth profile deviation in tooth height direction and helix deviation in tooth width direction. The helix deviation and tooth profile deviation of each gear tooth exist at the same time, so the accuracy evaluation of load distribution uniformity of gear should be considered comprehensively. When determining the gear tolerance, the tooth profile deviation is controlled by the tolerance item of gear transmission smoothness.

影响载荷分布均匀性的主要误差，在齿高方向是齿廓偏差，在齿宽方向是螺旋线偏差。齿轮每个轮齿的螺旋线偏差和齿廓偏差同时存在，齿轮载荷分布均匀性的精度评定应综合考虑。确定齿轮公差时,齿廓偏差由齿轮传动平稳性的公差项目来控制。

The main error that affects the rationality of backlash and the error that affects the size and unevenness of backlash is the tooth thickness deviation and its variation. Tooth thickness deviation refers to the difference between actual tooth thickness and nominal tooth thickness. The error of backlash is also related to the center distance of gear pair.

影响侧隙合理性的主要误差以及影响齿轮侧隙大小和侧隙不均匀的误差是齿轮的齿厚偏差及其变动量。齿厚偏差是指实际齿厚与公称齿厚之差。侧隙的误差还与齿轮副中心距有关。

6.1.3　Error Evaluation Items　误差评定项目

In the national standard, for the accuracy, stability and load distribution uniformity of gear transmission of the use requirements, the provisions of the accuracy index are stipulated.

国家标准中,对齿轮传动准确性、平稳性和载荷分布均匀性三个方面的使用要求,规定了检测精度指标。

In order to evaluate the accuracy of gear transmission, the national standard stipulates the compulsory inspection index of cumulative total pitch deviation, and sometimes it is necessary to increase the cumulative pitch deviation. The tangential comprehensive deviation, gear radial run-out and radial comprehensive total deviation are optional testing items.

为了评定齿轮传动准确性的精度,国标规定了强制性检测指标——齿距累积总偏差,有时还需要增加齿距累积偏差。切向综合总偏差、齿轮径向跳动和径向综合总偏差为非强制性检测项目。

In order to evaluate the accuracy of gear transmission smoothness, the national standard stipulates the compulsory inspection indexes of single pitch deviation and total deviation of tooth profile. One tooth tangential comprehensive deviation and one tooth radial comprehensive total deviation are optional testing items.

为了评定齿轮传动平稳性的精度,国标规定了强制性检测指标——单个齿距偏差和齿廓总偏差。一齿切向综合总偏差和一齿径向综合总偏差为非强制性检测项目。

In order to evaluate the accuracy of load distribution uniformity, the national standard stipulates the compulsory inspection index, which is the total deviation of tooth profile in the direction of tooth height and the total deviation of helix in the direction of tooth width.

为了评定载荷分布均匀性的精度,国标规定了强制性检测指标,在齿高方向为齿廓总偏差,在齿宽方向为螺旋线总偏差。

The gear pair backlash can be evaluated by tooth thickness deviation or common normal length deviation.

评定侧隙的指标,齿轮副侧隙大小可以用齿厚偏差或公法线长度偏差来评定。

6.2　Rolling Bearing　滚动轴承

Rolling bearing is a standard component widely used in modern industry. It is generally

Chapter 6 Tolerances and Fits for Typical Parts 典型零部件的公差与配合

composed of outer ring, inner ring, rolling feature and cage.

滚动轴承是现代工业中应用极其广泛的标准部件,一般由外圈、内圈、滚动体和保持架4部分组成。

According to the structural size, tolerance grade and technical performance of rolling bearing, the tolerance grade of rolling bearing is divided into five grades in *Rolling bearings—General technical regulations* (GB/T 307.3—2017). The tolerance grade of radial bearing is divided into five grades: 0, 6, 5, 4 and 2, while that of tapered roller bearing is 0, 6X, 5, 4 and 2, of which grade 0 is the lowest and level 2 is the highest. The tolerance grade of thrust bearing is divided into four grades: 0, 6, 5 and 4.

根据滚动轴承的结构尺寸、公差等级和技术性能,《滚动轴承　通用技术规则》(GB/T 307.3—2017)中将滚动轴承的公差等级分为五级。向心轴承的公差等级分为0、6、5、4、2五级,圆锥滚子轴承的公差等级分为0、6X、5、4、2五级。其中0级最低,2级最高。推力轴承的公差等级分为0、6、5、4四级。

Class 0 is a general level, which is used for general rotating mechanisms with low speed, medium speed and low rotation accuracy. Grade 6 is used for rotating mechanism with high speed and high precision. Grade 5 and level 4 are used for high-speed and high-precision mechanisms. Grade 2 is used for the mechanism with high speed and high precision.

0级为普通级,用于低速、中速和旋转精度要求不高的一般旋转机构。6级用于转速较高、旋转精度要求较高的旋转机构。5级、4级用于高速、高旋转精度要求的机构。2级用于转速很高、旋转精度要求也很高的机构。

In *Rolling bearings—Tolerances—Definitions* (GB/T 4199—2003), there are two kinds of provisions on the size tolerance of bearing inner diameter and outer diameter. The allowable deviation between the maximum and minimum values of the inner and outer diameter sizes, i.e. the single inner diameter and outer diameter deviation, is specified in order to limit the deformation. The average deviation of the maximum and minimum values of the actual size of the inner diameter and outer diameter measured in any cross section of the ring is specified, that is, the average inner diameter and outer diameter deviation of a single plane, so as to be used for bearing fitting.

《滚动轴承　公差　定义》(GB/T 4199—2003)中对轴承内径和外径尺寸公差做出了两种规定。规定了内径和外径尺寸的最大值和最小值所允许的偏差,即单一内径和外径偏差,目的是限制变形量。规定了套圈在任意横截面内测得的内径和外径实际尺寸的最大值和最小值的平均值偏差,即单一平面平均内径和外径偏差,目的是用于轴承的配合。

The tolerance zone of inner diameter and outer diameter of radial bearing is unidirectional. The tolerance zone is located below the zero line with the inner diameter and outer diameter as the zero line, that is, the upper limit deviation is zero and the lower limit deviation is negative. The fit of bearing inner ring and journal is base hole system, and that of bearing outer ring and shell hole is base shaft system.

向心轴承内径和外径的公差带均为单向制,统一采用公差带位于以内径和外径为零线的下方,即上极限偏差为0,下极限偏差为负值。轴承内圈与轴颈的配合为基孔制,轴承外圈与外壳孔的配合为基轴制。

Rolling bearing is a standard part. The tolerance zone of bearing inner ring diameter and outer ring journal has been determined during manufacturing. Therefore, the fit of bearing with journal and housing hole needs to be determined by the tolerance zone of journal and housing hole.

滚动轴承是标准件,轴承内圈孔径和外圈轴径公差带在制造时已经确定,因此轴承与轴颈和轴承座孔的配合需要由轴颈和轴承座孔的公差带决定。

The correct and reasonable selection of the fitting of the rolling bearing with the journal and the shell hole has a great influence on ensuring the normal operation of the machine, improving the service life of the bearing and giving full play to its bearing capacity. Therefore, the selection of tolerance zone between journal and housing hole should be based on the type, size, radial clearance and working conditions of the load acting on the bearing. At the same time, the working temperature, structure and material of journal and shell hole, rotation accuracy and speed, bearing installation and disassembly are considered.

正确、合理地选用滚动轴承与轴颈和轴承座孔的配合,对保证机器正常运转、提高轴承的使用寿命、充分发挥其承载能力影响很大。因此,选用轴颈与轴承座孔公差带时,要以作用在轴承上负荷的类型、大小、径向游隙和工作条件为依据。同时考虑工作温度、轴颈与轴承座孔的结构和材料、旋转精度与速度、轴承的安装与拆卸情况等因素。

6.3　Key Connection　键连接

The key connection is usually used as a detachable connection between the shaft and the shaft to transfer torque or to guide the shaft upper part. Key connection is generally divided into single key connection and spline connection. Single key can be divided into flat key, semicircular key and wedge key, and spline can be divided into rectangular spline and involute spline. Here, only the interchangeability of flat key and rectangular spline connection are briefly discussed.

键连接通常用作轴和轴上传动件的可拆卸连接,用以传递转矩或用作轴上传动件的导向。键连接一般分为单键连接和花键连接两类,其中单键可分为平键、半圆键和楔键,花键可分为矩形花键和渐开线花键。这里,只简单讨论平键和矩形花键连接的互换性。

For the interchangeability of flat key connection, it should be ensured that the side of key and keyway should have sufficient effective contact area to bear the load, so as to ensure the strength, service life and reliability of key connection. In the parallel key connection, the key width and the keyway width are the fitting sizes. The key width can be regarded as the shaft and the keyway width can be regarded as the hole. Therefore, the fitting system of key width and keyway width should adopt the base shaft system, which can meet the fitting performance requirements of different key connection through different keyway width tolerance.

平键连接的互换性要求:应保证键与键槽的侧面有充分的有效接触面积来承受负荷,并保证键连接的强度、寿命和可靠性。平键连接中,键宽和键槽宽是配合尺寸,键宽可以视为轴,键槽宽可以视为孔。因此键宽与键槽宽的配合制应采用基轴制,通过不同的键槽宽公差带来满足不同键连接的配合性能要求。

Spline connection is the combination of spline shaft and spline hole, which can be used as

Chapter 6 Tolerances and Fits for Typical Parts 典型零部件的公差与配合

fixed connection or sliding connection. According to the national standard, the small diameter centering is adopted for the connection of rectangular spline. In order to reduce the variety and specification of broach and measuring tool used in manufacturing internal spline, the rectangular spline should be matched with the base hole system. The limit and fit selection of rectangular spline connection is mainly to determine the connection accuracy and assembly form. In order to ensure the fitting property of the centering surface, the dimensional tolerance and geometric tolerance of the centering diameter of the internal and external splines need to adopt the inclusion requirements. The positional tolerance of the side to the centering shaft shall be specified for the spline, and the maximum material requirement shall be adopted. For the single piece and the small batch production, the symmetry tolerance of the central plane of key or keyway to the centering shaft should be specified, and the opposite principle should be adopted. The surface roughness parameter of rectangular spline is generally marked with the upper limit value of Ra.

花键连接是花键轴与花键孔两个零件的结合,可用作固定连接,也可用作滑动连接。国家标准规定,矩形花键连接采用小径定心。为了减少制造内花键用的拉刀和量具的品种规格,矩形花键配合应采用基孔制。矩形花键连接的公差与配合选用主要是确定连接精度和装配形式。为保证定心表面的配合性质,内、外花键定心直径的尺寸公差与几何公差需要采用包容要求,花键应规定侧面对定心轴线的位置度公差并采用最大实体要求。单件小批量生产,应规定键或键槽中心平面对定心轴线的对称度公差,并采用独立原则。矩形花键的表面粗糙度参数一般标注 Ra 的上限值。

Exercises 6 习题 6

6-1　What are the requirements of gear transmission?

6-1　齿轮传动有哪些使用要求?

6-2　What are the main error sources that affect the accuracy of gears?

6-2　影响齿轮精度的主要误差来源是什么?

6-3　What are the tolerance grades of rolling bearings? Which grade is used most commonly?

6-3　滚动轴承的公差等级有几个? 用得最多的是哪个等级?

6-4　What standard system is adopted for the fit of inner ring and shaft, outer ring and housing hole of rolling bearing?

6-4　滚动轴承内圈与轴、外圈与外壳孔的配合分别采用哪种基准制?

References 参考文献

[1] 中国国家标准化管理委员会. 产品几何技术规范(GPS) 技术产品文件中表面结构的表示法:GB/T 131—2006 [S]. 北京:中国标准出版社, 2006.

[2] 中国国家标准化管理委员会. 产品几何技术规范(GPS) 几何公差 形状、方向、位置和跳动公差标注:GB/T 1182—2018 [S]. 北京:中国标准出版社, 2018.

[3] 中国国家标准化管理委员会. 产品几何技术规范(GPS) 几何公差 检测与验证:GB/T 1958—2017 [S]. 北京:中国标准出版社, 2017.

[4] 中国国家标准化管理委员会. 产品几何技术规范(GPS) 基础 概念、原则和规则:GB/T 4249—2018 [S]. 北京:中国标准出版社, 2018.

[5] 中国国家标准化管理委员会. 产品几何技术规范(GPS) 几何公差 最大实体要求(MMR)、最小实体要求(LMR)和可逆要求(RPR):GB/T 16671—2018 [S]. 北京:中国标准出版社, 2018.

[6] 中国国家标准化管理委员会. 产品几何技术规范(GPS) 几何公差 轮廓度公差标注:GB/T 17852—2018 [S]. 北京:中国标准出版社, 2018.

[7] 张彦富. 几何量公差与测量技术基础(英文版)[M]. 北京:北京航空航天大学出版社, 2015.

[8] 宋绪丁. 互换性与几何量测量技术[M]. 3版. 西安:西安电子科技大学出版社, 2019.

[9] 周兆元. 互换性与测量技术基础[M]. 4版. 北京:机械工业出版社, 2019.

[10] 王宏宇. 互换性与测量技术[M]. 北京:机械工业出版社, 2019.

[11] 于雪梅. 互换性与技术测量 [M]. 北京:机械工业出版社, 2020.